Slow Ways Pocket Atlas

Slow Ways supporters
Slow Ways CIC

Words by
Hannah Engelkamp
Daniel Raven-Ellison

Maps by
Charlie Peel

Slow Ways

Published by

urban good

October 2024
Slow Ways Pocket Atlas
Words by Hannah Engelkamp and Daniel Raven-Ellison
Slow Ways is being created by thousands of volunteers
Maps & design by Charlie Peel, Urban Good
www.slowways.org
hello@slowways.org

Published by Urban Good CIC
2 Grosvenor Way, London, E5 9ND
www.urbangood.org
Design © 2024 Urban Good CIC

Printed in Great Britain by Generation Press – a carbon neutral B Corp
Carbon balanced paper is Ultra Fine Offset by Denmaur
This book is set in Akkurat by Lineto

We believe it should be possible to travel all of the connections in this atlas safely and enjoyably but that might not be the case yet. Enjoy the routes at your own risk. Content is based on mapping data compiled in 2023 and 2024. Travellers are advised to check the Slow Ways website for up-to date routes and journey details.

Thank you to Becky Duncan, Jessie Leong, Katie McGoran, Roxy Barry, Hannah Engelkamp, Mar Davenport, David Lintern, Ingrina Shieh and Phil Young for photography throughout this atlas.

All of the maps in this book also exist in Welsh and Scottish Gaelic, in a folded map format. Thes are available to order from your local bookshop or from www.urbangood.org.

With deep thanks to the many thousands of people who have contributed to Slow Ways so far.

Slow Ways would not exist without you.

ESEA Outdoors (East and Southeast Asian outdoor enthusiasts) walking Slow Ways from Bristol to Gloucester to Oxford

Contents

Human beings are walking creatures

We got ourselves all over the world, learned to read landscapes, found the ways through mountain passes and across rivers. We were curious, restless, adventurous and versatile. We solved problems by keeping mobile.

And we were – and are – social. We have always asked each other where to go. The paths we walked created the roads we still travel. We shared information about the crossing places, sheltered places, abundant places. These meeting places became settlements. Many of these settlements are still here.

Whether travelling for joy, love, work, safety, reflection, opportunity or adventure, navigating across land is shown to actually create the neural pathways in our brains. The routes we travel were created by our ancestors' brains, but they also create our future brains.

These routes are our birthright.

But the best way may now be unknown, untrodden or non-existent. Where old ways are now A-roads or built over, or where there is no safe way or the land is now private, or there are new important places to reach, we are finding new ways to go. The routes we seek to create have all the same priorities as they would have had for our ancestors – we want them to be direct, safe, enjoyable and as easy to travel as possible.

Emerging is a network of trust and advice that is creating ease and confidence. Information about routes is making it possible for people to decide what they can do, rather than waiting to be told. Each walk is recounted in a review that gives guidance and tells a story of the journey. Footstep upon footstep, story upon story, tales upon trails upon tales, intertwined, collaborative, recreating the relationship with the landscape that we had once and will have again.

Our Slow Ways network is an inheritance, a gift for us in the present and a legacy for future generations.

Slow Ways

This pocket atlas shows you where all the Slow Ways go. It is designed for planning journeys and colouring in where you have been.

It should be used as a companion to the Slow Ways website at www.slowways.org.

Bristol Steppin Sistas on Bristol to Wick

So what are Slow Ways?

A town-to-town walking and wheeling network

Slow Ways routes are citizen-sourced suggestions for the best way to walk, or wheel where possible, between any town or city and its neighbours.

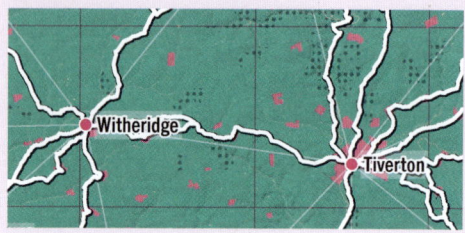

Sharing the best ways

The network is designed around the Slow Ways principle that there should always be a good way to walk or wheel between neighbouring settlements. Some Slow Ways have one route option. Others have multiple route options that will be the best choices for different people, landscapes or occasions.

Suggested, tested and improved by real people

Computer-algorithm-automated routes have their place. But Slow Ways routes have been suggested by people, and have then been walked or wheeled and experienced by other people. People have explored routes they'd not have risked without the guidance from people who've passed that way before.

Covering the whole of Great Britain

The whole network already exists. It consists of over 9,000 routes that connect over 2,500 settlements. All of the routes make use of existing paths, trails, rights of way and accessible areas. We have plans to cover Ireland soon too.

Peer reviews that offer guidance

On the Slow Ways website each walk is recounted in a review that gives guidance and tells a story of the journey. Some 1.5 million words have been shared so far – that's roughly 20 books' worth of inspiration and advice.

Using tech to reconstruct the knowledge we once had
Humanity's oldest infrastructure and its youngest
technologies are working together – we are using
online mapping and reviews to assist people in telling
each other the best way to go. By providing these tools
a network of trust and guidance is emerging, that is
creating ease and confidence for the next walkers or
wheelers.

**Making the information available enables people to
make better decisions**
Through gathering information about routes – especially
path condition and obstacles – it becomes possible for
people with all sorts of access needs to make up their
own minds about what they can do.

And the network needs you
The routes will always need new reviews and fresh
checks. Consider this pocket atlas an invitation to join in
and enjoy.

Frankie and Frit on a route in Sheffield

The future

We deserve and need an extensive and easy-to-use walking and wheeling network. That is what we're co-creating, and by exploring this atlas you are already part of Slow Ways' future.

Good walking and wheeling routes are pathways to longer and richer lives.

The benefits of delivering such a network are numerous, and the consequences of not making it easy for people to enjoy walking and wheeling journeys is already too costly for individuals and our communities, public services, nation, planet and climate.

Walking is good for the mind and body. It draws you into being part of the natural world. It creates practical and emotional connections between people and communities. As a pastime it's economical – both for your own pocket and the planet's limited resources. Where routes exist there is something to protect and improve. And as the network grows it becomes a superhighway for ideas, memories, friendships and trust.

While Britain is rich with paths and rights of way, we can do much better in how they are mapped, understood and made use of.

Our paths are by no means as used, extensive, inclusive, signed, managed or funded as they should be. And we don't just need to improve our existing paths – we need new ones too.

We need a walking and wheeling network for:

everyone

Our future walking and wheeling network should be as safe, inclusive and accessible as possible so that it can benefit as many people as possible.

And access and connectivity are not just important for humans. They are critical for the rest of nature as well. Our future network should be designed to help wildlife move through the landscape too.

everywhere

There should be a good way to walk or wheel to any address in Britain. As a rule of thumb, if you can drive somewhere you should also be able to travel there under your own steam.

And where there is a right to roam, we should share recommended ways to go that are sensitive to people, place, land and life.

everyday

Walking and wheeling should be part of everyday life. People should be able to walk or wheel to see friends, go to the shops, get to school or immerse themselves in nature.

We don't even know what a country with a fully tested, trusted walking and wheeling network would feel like to live in. How will it develop? What will it enable? What will it mean to people?

We are not waiting for governments to make this happen. Thousands of people are excited by this great step forward, and are helping to bring it about. Sheer citizen-power is making this a reality.

We call on our local and national governments to get involved too! A proper national walking and wheeling network would be popular with the public and a great investment in the future.

In the meantime, most of the routes in this atlas are ready to be enjoyed – just not by as many people as we would like. That is not good enough.

By being part of this national effort we can share good routes, bad routes and where new routes are needed. That information can then help out people who couldn't walk or wheel without it.

Enjoy the following 114 pages of maps – they are a lifetime and more of possibilities!

Turn to the back for an easy step-by-step guide to using, contributing to and make the most of Slow Ways.

Slow Ways create ways, time, space and infrastructure for joy, memory, relationships, hope, connection, community, story, adventure, imagination, generosity, health, travel, creativity, climate action, learning, wellbeing, legacy, freedom, expression, collaboration, possibility and more... all by going for walks.

A Slow Way from Rutherglen to East Kilbride

People colouring in the routes they've walked

Maps

Coverage

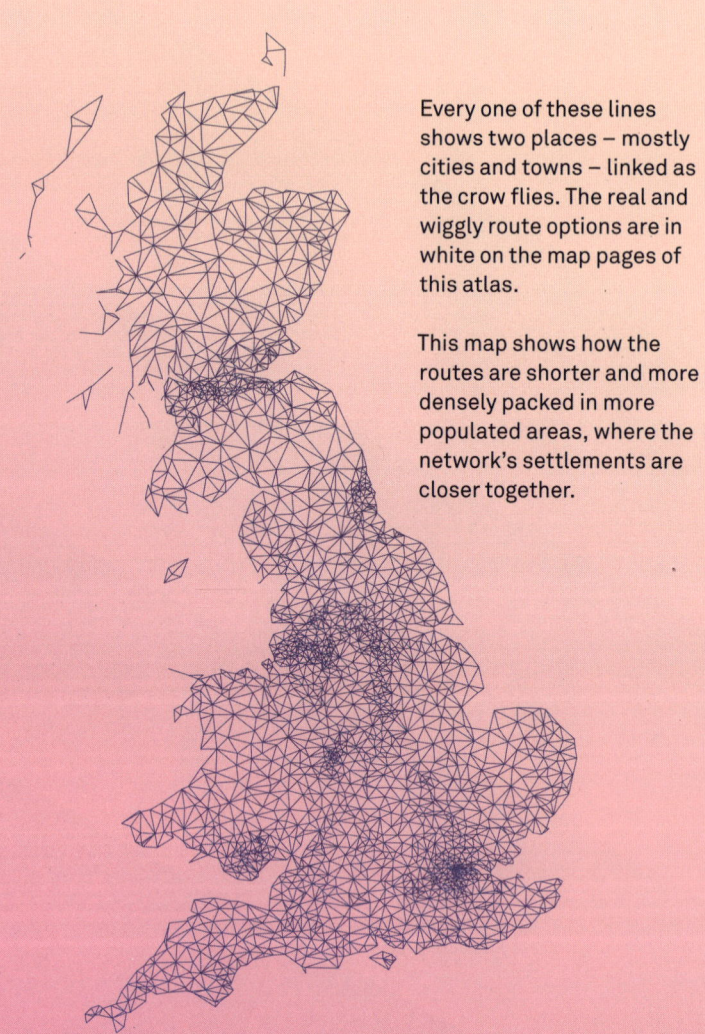

Every one of these lines shows two places – mostly cities and towns – linked as the crow flies. The real and wiggly route options are in white on the map pages of this atlas.

This map shows how the routes are shorter and more densely packed in more populated areas, where the network's settlements are closer together.

Page finder

140 – 141

138 – 139

132 – 133 | 134 – 135 | 136 – 137

124 – 125 | 126 – 127 | 128 – 129 | 130 – 131

INVERNESS / INBHIR NIS

116 – 117 | 118 – 119 | 120 – 121 | 122 – 123

CAIRNGORMS | ABERDEEN

108 – 109 | 110 – 111 | 112 – 113 | 114 – 115

TAY FOREST PARK | DUNDEE
LOCH LOMOND & THE TROSSACHS | PERTH

100 – 101 | 102 – 103 | 104 – 105 | 106 – 107

STIRLING | EDINBURGH
GLASGOW

94 – 95 | 96 – 97 | 98 – 99

GALLOWAY FOREST PARK | NORTHUMBERLAND

86 – 87 | 88 – 89 | 90 – 91 | 92 – 93

CARLISLE | NEWCASTLE SUNDERLAND DURHAM
LAKE DISTRICT | NORTH YORK MOORS

80 – 81 | 82 – 83 | 84 – 85

LANCASTER | YORKSHIRE DALES RIPON | YORK
PRESTON | BRADFORD LEEDS WAKEFIELD | KINGSTON UPON HULL

72 – 73 | 74 – 75 | 76 – 77 | 78 – 79

BANGOR | LIVERPOOL MANCHESTER | SALFORD SHEFFIELD | LINCOLN
ST ASAPH CHESTER | PEAK DISTRICT

62 – 63 | 64 – 65 | 66 – 67 | 68 – 69 | 70 – 71

SNOWDONIA | STOKE-ON-TRENT NOTTINGHAM DERBY | THE BROADS
WOLVERHAMPTON BIRMINGHAM | LEICESTER PETERBOROUGH ELY | NORWICH

52 – 53 | 54 – 55 | 56 – 57 | 58 – 59 | 60 – 61

ST DAVIDS | HEREFORD | WORCESTER COVENTRY | CAMBRIDGE
BRECON BEACONS | GLOUCESTER

PEMBROKESHIRE COAST

42 – 43 | 44 – 45 | 46 – 47 | 48 – 49 | 50 – 51

SWANSEA/ABERTAWE | NEWPORT/CASNEWYDD | OXFORD | ST ALBANS | CHELMSFORD
EXMOOR | CARDIFF/CAERDYDD WELLS | BRISTOL BATH | LONDON | CANTERBURY

32 – 33 | 34 – 35 | 36 – 37 | 38 – 39 | 40 – 41

EXETER | WINCHESTER SALISBURY SOUTHAMPTON | SOUTH DOWNS
DARTMOOR | NEW FOREST | PORTSMOUTH BRIGHTON

28 – 29 | 30 – 31

TRURO | PLYMOUTH

Reading the maps

Points
- City
- Town / hub

Lines
- Slow Way
- Route option
- Rail
- National Trail / Scotland's Great Trail / Wales Coast Path

Areas
- Land
- Built-up area
- Water
- Wooded area
- National Park
- National Landscape / National Scenic Area (Scotland)

Scale
The maps are drawn at 1:500,000.
This means that 1cm of map is 5km in the real world.
Our 10km grid (6.2 miles) helps to quickly estimate distances.

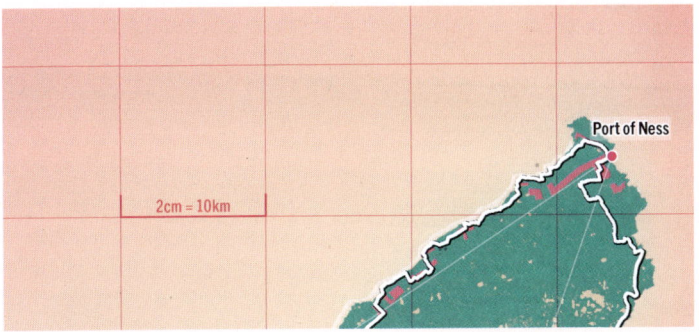

Keep a record
You can colour in the routes or the straight, white network lines to plan your journeying, or to tick off the ones you have completed.

Page turners
Within the white box on every map spread you will find the corresponding page number to turn to when your route goes off the page. There is 15mm of overlap on every page of maps, so you can easily pick up where you left off.

Map data and copyright
Slow Ways CIC, Slow Ways Contributors 2024.
Contains OS data © Crown copyright 2024.
Contains OpenStreetMap data © OpenStreetMap contributors 2024.
Design © Charlie Peel, Urban Good CIC, 2024.

May your day be full of magic and
may you not be too busy to see it.
Brad Montague

St Ives

Treen

Hayle

St Just

Penzance

Marazion

Newlyn

Land's End

SOUTH WES
COAST PATH

Porthcurno

CORNWALL

Bolventor

↑33

Pensilva

Callington

Tavistock

Calstock

Princetown

Yelverton

Bodmin

Lanivet

Liskeard

TAMAR
VALLEY

ugle

Lostwithiel

Penwithick
Blazey

Saltash

Austell

Looe

Torpoint

PLYMOUTH

Plymstock

Fowey

SOUTH WEST
COAST PATH

evagissey

↑29

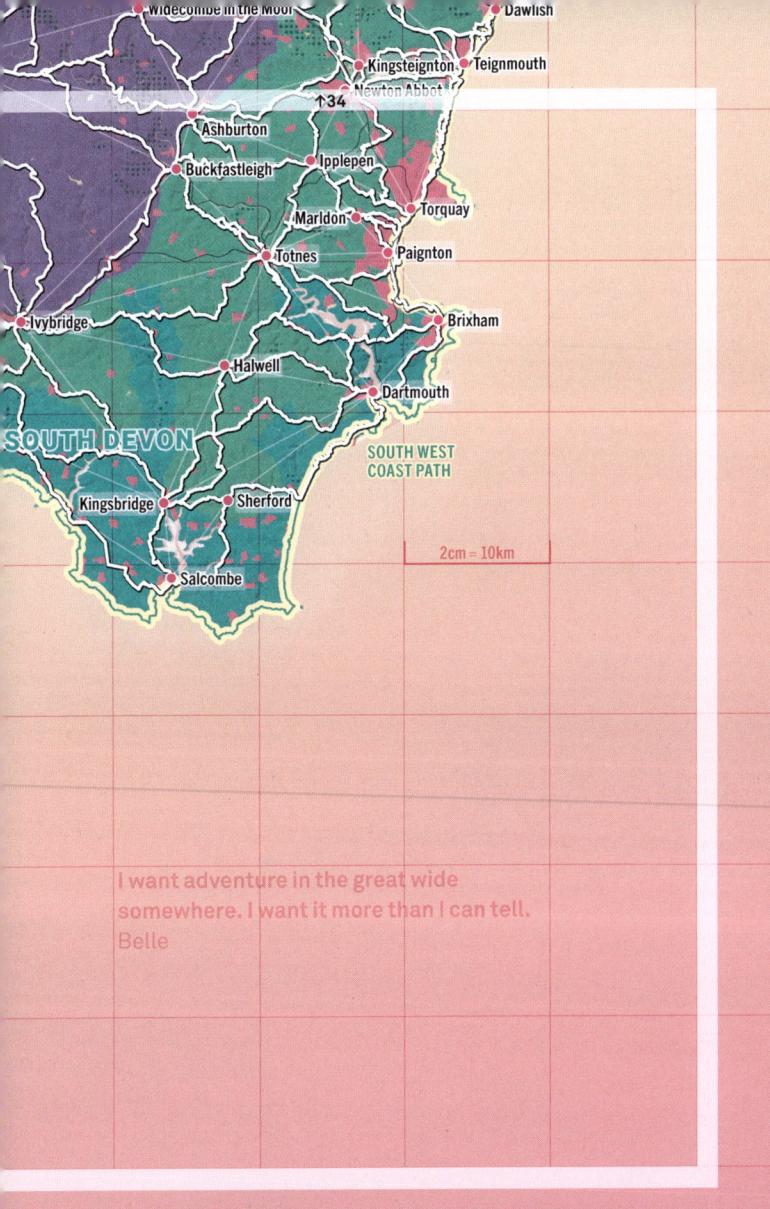

Widecombe in the Moor · Dawlish

Kingsteignton · Teignmouth

↑34 Newton Abbot

Ashburton

Buckfastleigh · Ipplepen

Marldon · Torquay

Totnes · Paignton

Ivybridge · Brixham

Halwell

Dartmouth

SOUTH DEVON

SOUTH WEST
COAST PATH

Kingsbridge · Sherford

2cm = 10km

Salcombe

I want adventure in the great wide
somewhere. I want it more than I can tell.
Belle

↑42

Never doubt that a small group of thoughtful, committed citizens can change the world; indeed, it is the only thing that ever has.
Margaret Mead

Tintagel

SOUTH WEST
COAST PATH

2cm = 10km

Port Isaac

Padstow

Wadebridge

↓29

Bodmin

Lanivet

St Columb Major

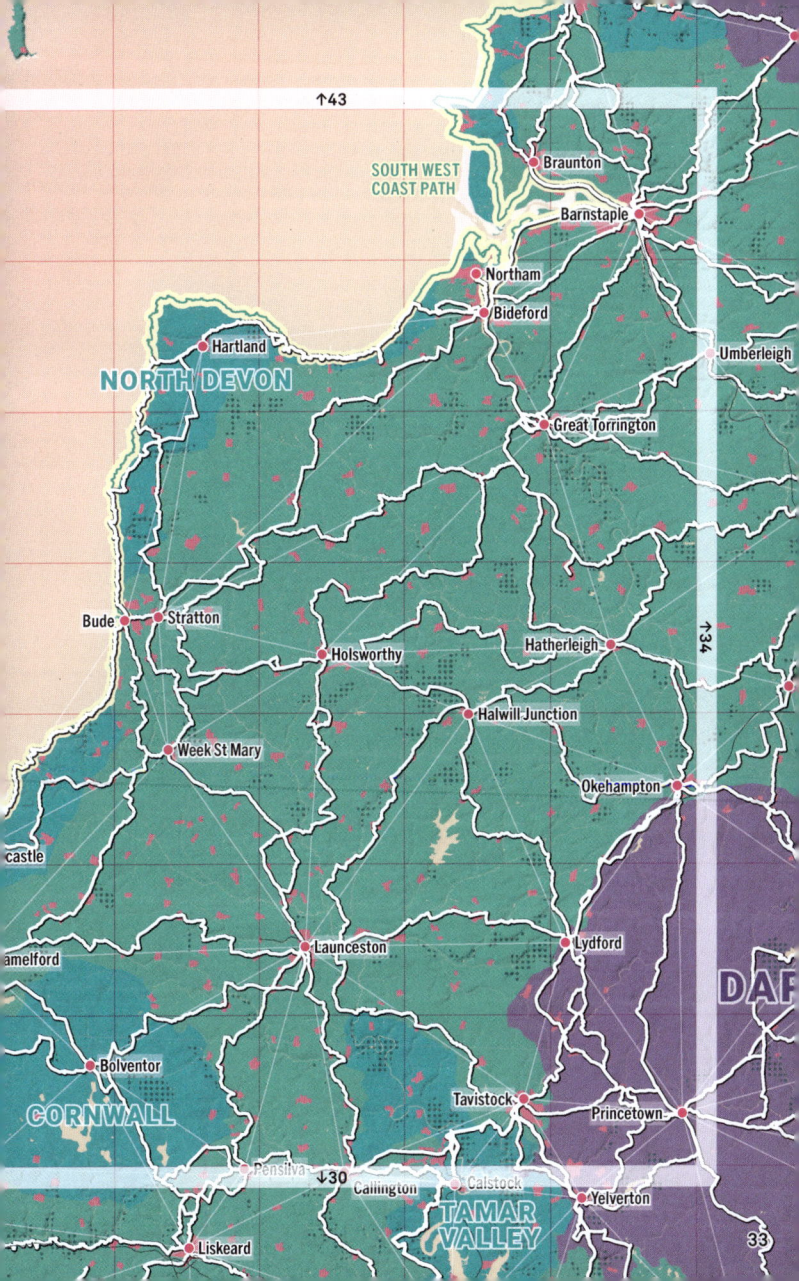

SOUTH WEST
COAST PATH

Braunton

Barnstaple

Northam

Bideford

Umberleigh

Hartland

NORTH DEVON

Great Torrington

Bude

Stratton

Holsworthy

Hatherleigh

Halwill Junction

Week St Mary

Okehampton

castle

amelford

Launceston

Lydford

DAF

Bolventor

Tavistock

Princetown

CORNWALL

Pensilva

Callington

Calstock

Yelverton

**TAMAR
VALLEY**

Liskeard

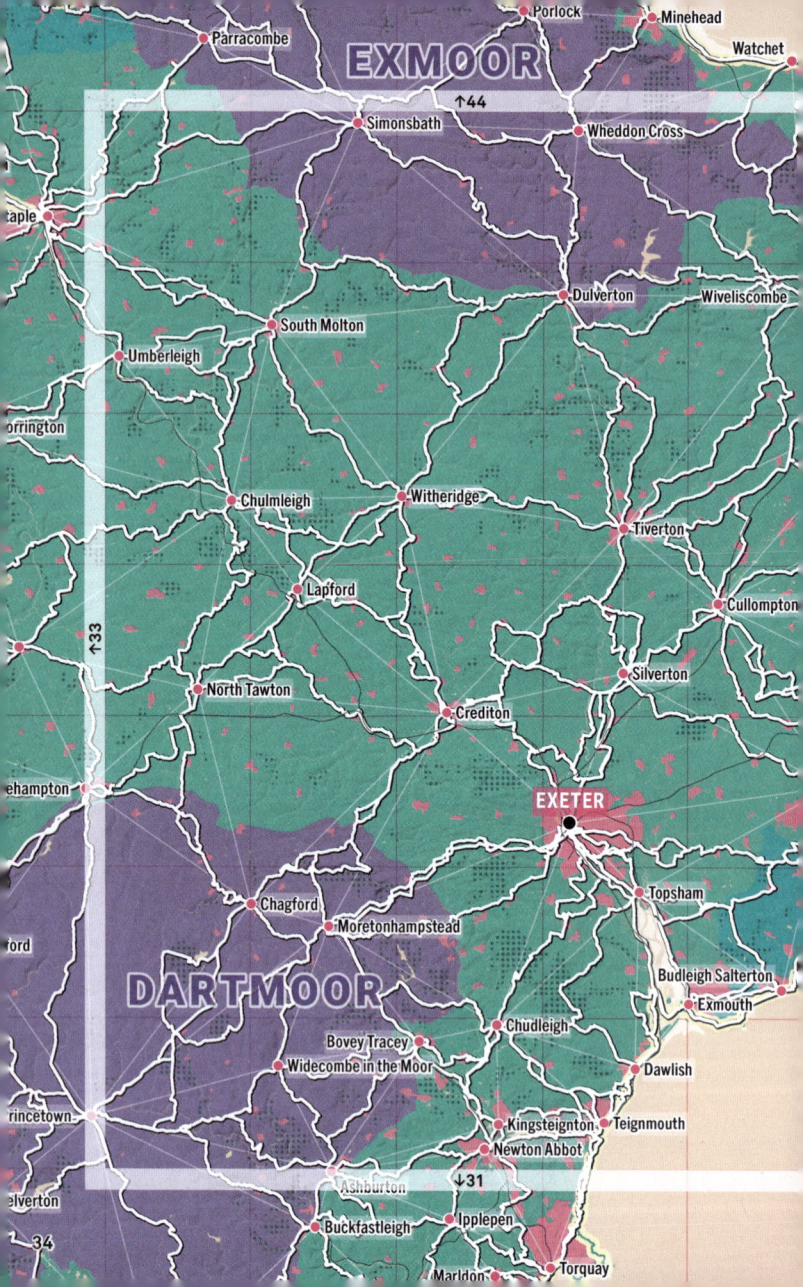

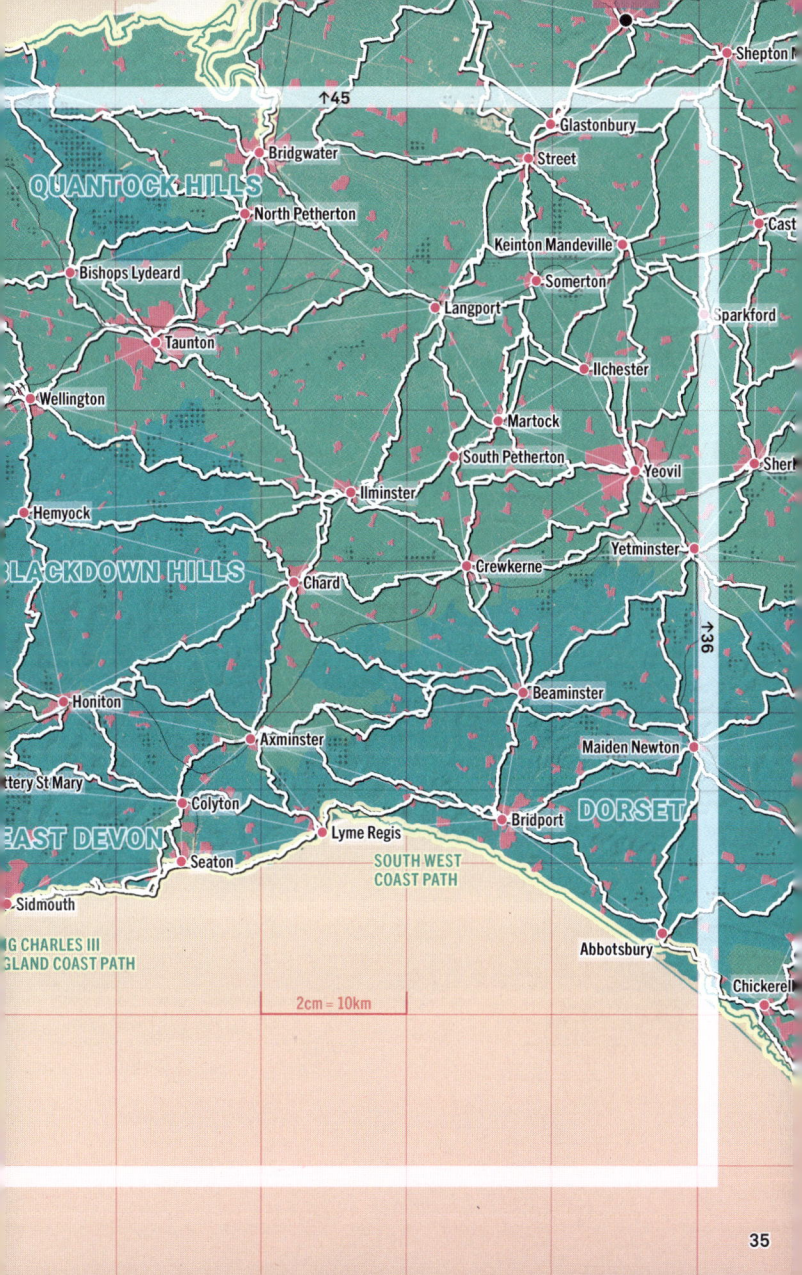

QUANTOCK HILLS

Bridgwater

North Petherton

Bishops Lydeard

↑45

Taunton

Wellington

Hemyock

BLACKDOWN HILLS

Honiton

ttery St Mary

EAST DEVON

Colyton

Seaton

Sidmouth

NG CHARLES III
GLAND COAST PATH

Glastonbury

Street

Keinton Mandeville

Cast

Somerton

Sparkford

Langport

Ilchester

Martock

South Petherton

Yeovil

Sher

Ilminster

Yetminster

Chard

Crewkerne

Axminster

Beaminster

Maiden Newton

DORSET

Bridport

Lyme Regis

SOUTH WEST
COAST PATH

Abbotsbury

Chickerell

2cm = 10km

↑36

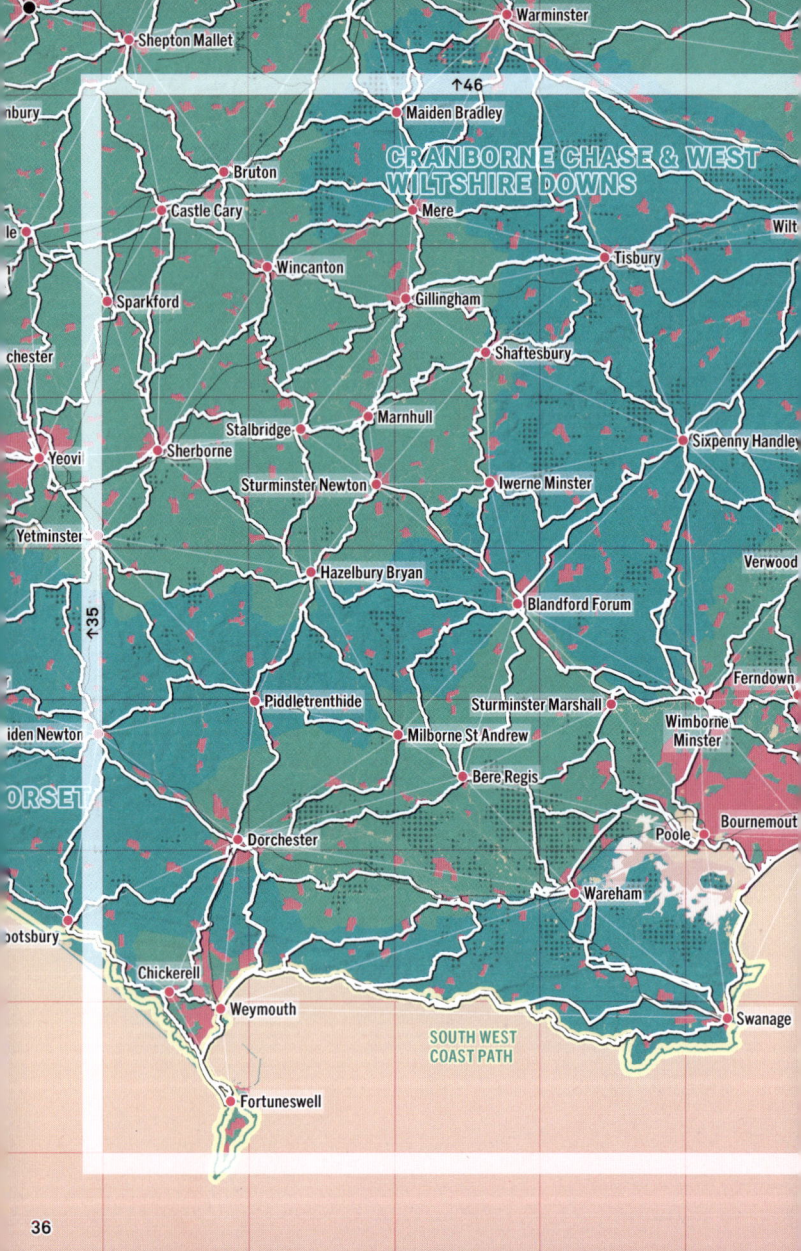

Shepton Mallet

Warminster

↑46

Maiden Bradley

CRANBORNE CHASE & WEST
WILTSHIRE DOWNS

nbury

Bruton

Castle Cary

Mere

Tisbury

Wilt

le

Wincanton

Gillingham

Sparkford

chester

Shaftesbury

Stalbridge

Marnhull

Sixpenny Handley

Yeovil

Sherborne

Sturminster Newton

Iwerne Minster

Yetminster

Verwood

Hazelbury Bryan

↑35

Blandford Forum

Ferndown

Piddletrenthide

Sturminster Marshall

iden Newton

Milborne St Andrew

Wimborne
Minster

ORSET

Bere Regis

Bournemout

Poole

Dorchester

Wareham

ootsbury

Chickerell

Weymouth

Swanage

SOUTH WEST
COAST PATH

Fortuneswell

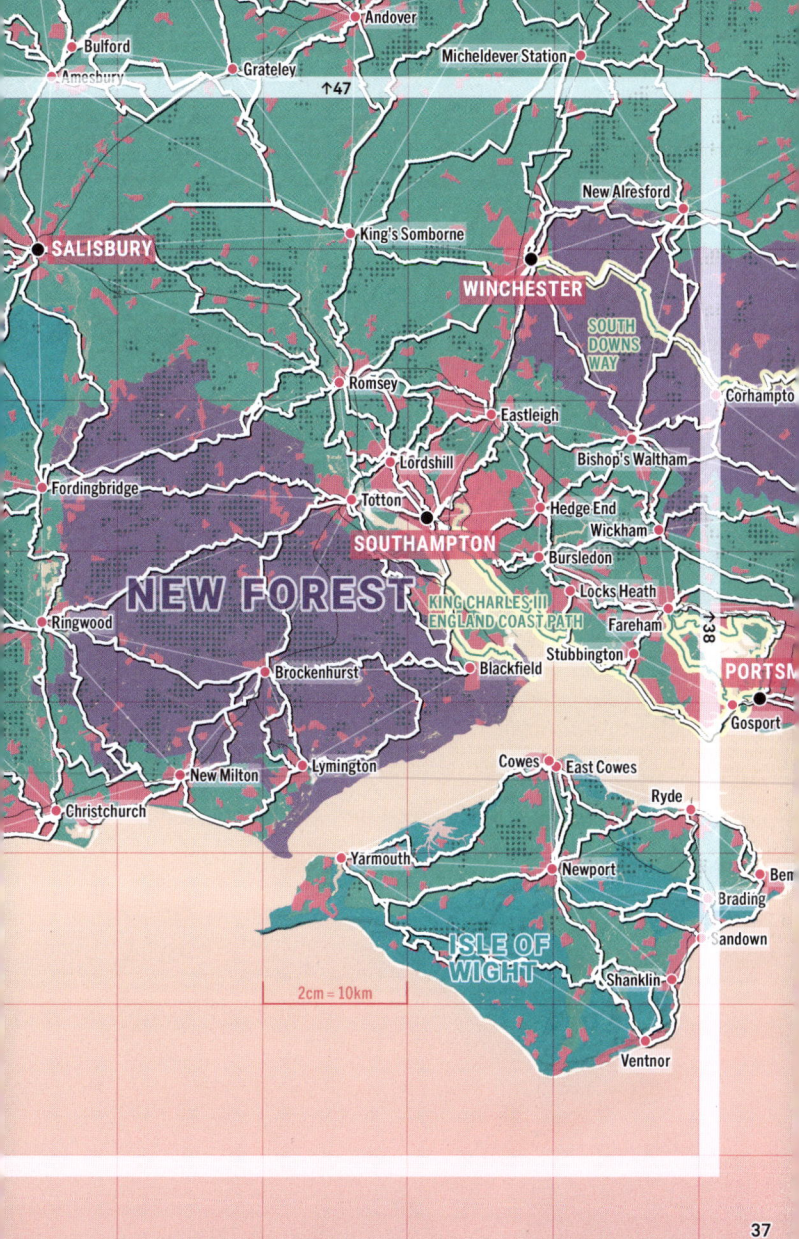

Bulford
Amesbury
Andover
Grateley
Micheldever Station
↑47
SALISBURY
New Alresford
King's Somborne
WINCHESTER
SOUTH
DOWNS
WAY
Corhampto
Romsey
Eastleigh
Bishop's Waltham
Fordingbridge
Lordshill
Totton
Hedge End
Wickham
SOUTHAMPTON
Bursledon
NEW FOREST
KING CHARLES III
ENGLAND COAST PATH
Locks Heath
Fareham
↑38
Ringwood
Brockenhurst
Blackfield
Stubbington
PORTSM
Christchurch
New Milton
Lymington
Cowes
East Cowes
Ryde
Gosport
Yarmouth
Newport
Ben
Brading
Sandown
ISLE OF
WIGHT
Shanklin
2cm = 10km
Ventnor

37

SURREY HILLS

Farnham
Bentley
Godalming
↑48
Alton
Cranleigh
w Alresford
Bordon
Haslemere
Loxwoo
Liphook
Billingshurst
SOUTH DOWNS
UTH
OWNS
AY
Petersfield
Midhurst
Petworth
Corhampton
West Chiltington Common
op's Waltham
Storrington
nd
Horndean
ckham
n
ks Heath
Havant
Arundel
Fareham ↑37
Emsworth
CHICHESTER
Westergate
ton
PORTSMOUTH
CHICHESTER
Littlehampton
HARBOUR
Gosport
Hayling
Bognor Regis
owes
KING CHARLES III
ENGLAND COAST PATH
Ryde
Selsey
rt
2cm = 10km
Bembridge
Brading
Sandown
Shanklin
Ventnor

38

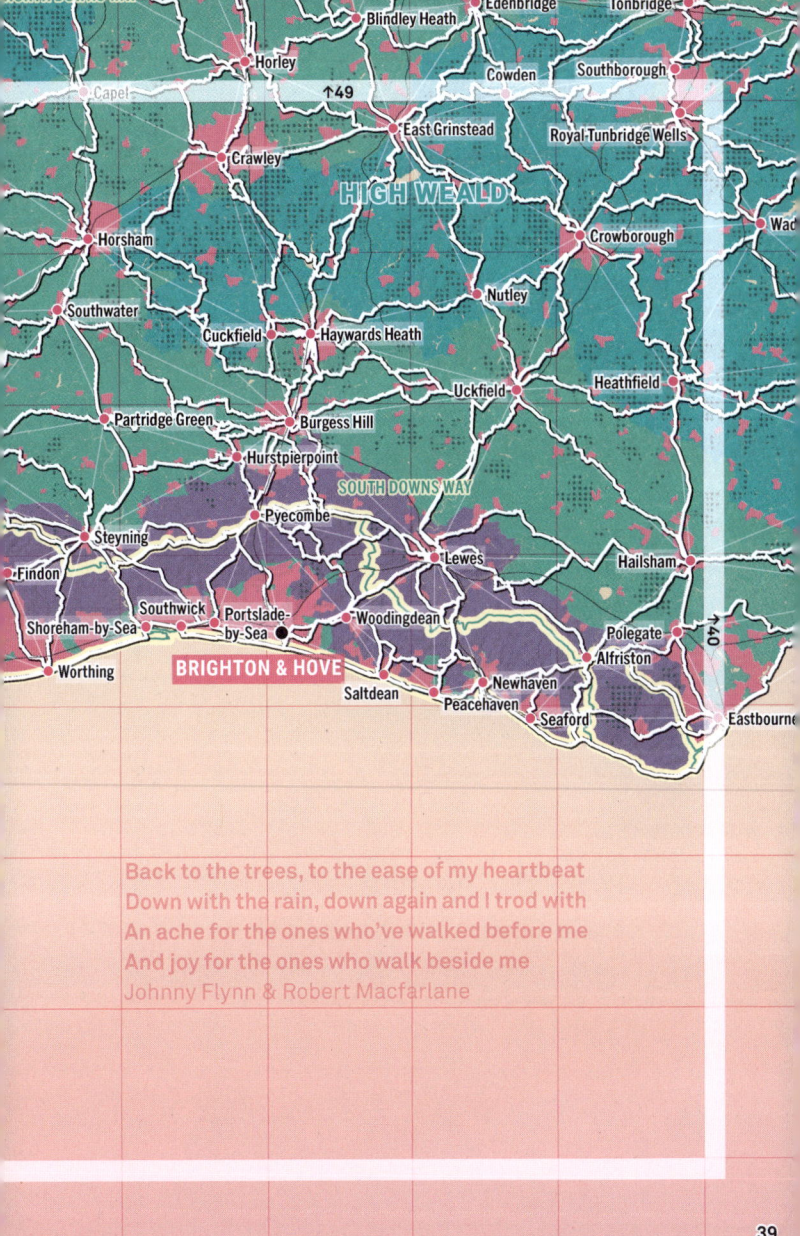

Back to the trees, to the ease of my heartbeat
Down with the rain, down again and I trod with
An ache for the ones who've walked before me
And joy for the ones who walk beside me
Johnny Flynn & Robert Macfarlane

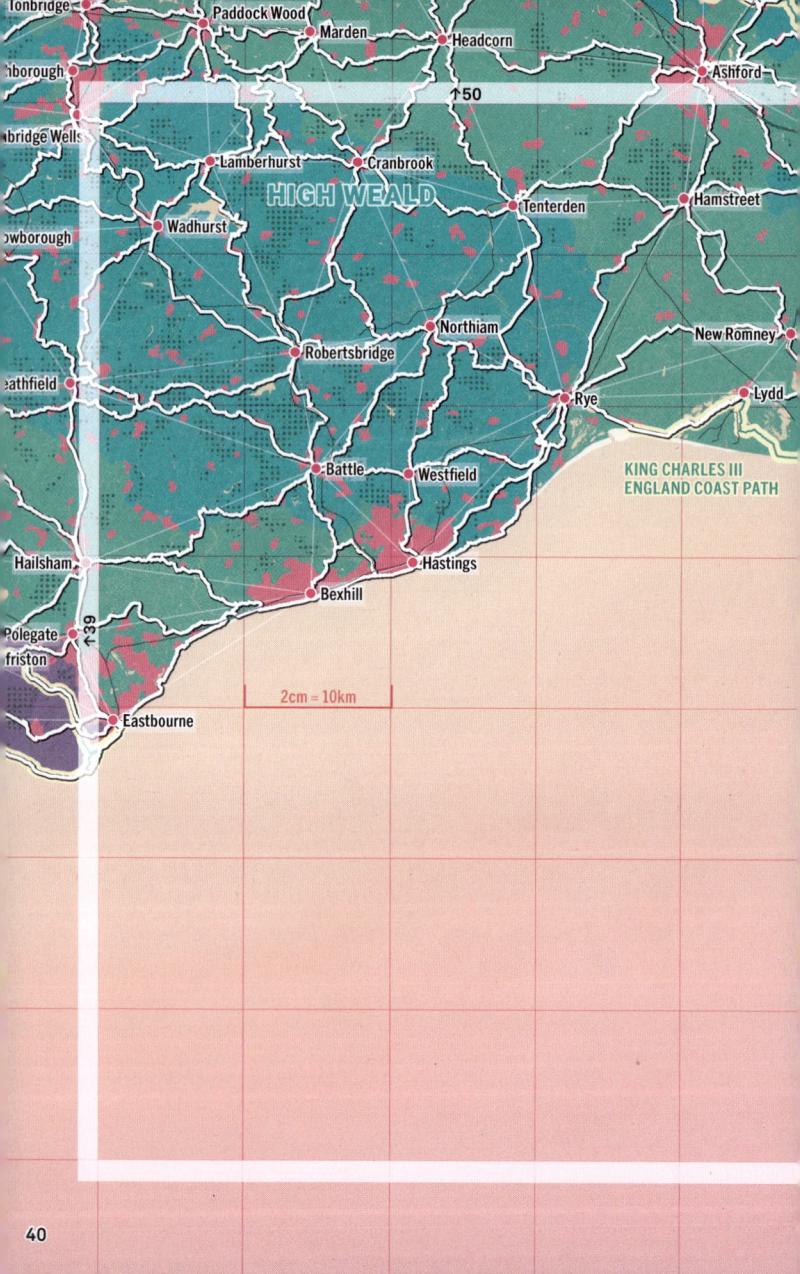

Tonbridge

Paddock Wood

Marden

Headcorn

Ashford

...borough

↑50

...bridge Wells

Lamberhurst

Cranbrook

Hamstreet

HIGH WEALD

Tenterden

Wadhurst

...owborough

New Romney

Northiam

Robertsbridge

Rye

Lydd

...eathfield

Battle

Westfield

KING CHARLES III
ENGLAND COAST PATH

Hailsham

Hastings

↑39

Polegate

Bexhill

...friston

2cm = 10km

Eastbourne

KENT DOWNS

Lyminge
Dover
↑51
Hawkinge
Folkestone
Hythe

Slow down. Life is crossing the road.
Debasish Mridha

41

Haverfordwest

Milford Haven Neyland

Pembroke Dock

Pembroke

The question is: what
happens after you walk
off into the sunset?
Bridget Jones

I didn't know where I was going until I got there.
Christopher McCandless

I live my life a quarter mile at a time.
Dominic Toretto

WALES COAST PATH

2cm = 10km

SOUTH WEST COAST PATH

KING CHARLES III ENGLAND COAST PATH

EXMOOR

Ammanford
Brynamman
Ystradfellte
Nant-ddu
yberem
Ystradgynlais ↑54
Glyn-Neath
Merthyr Tydfil
ntarddulais
Pontardawe
Crynant
Hirwaun
Aberdare
Gorseinon
Mountain Ash
Loughor
Neath
Treherbert
Ferndale
Treha
Gowerton
Treorchy
Tylorstown
Abércyno
SWANSEA/ABERTAWE
Tonypandy
Porth
Port Talbot
Maesteg
Pontycymer
Pontypridd
Tonyrefail
Church Villag
Beddau
Sarn
Llantrisant
Pyle
Pencoed
Talbot Green
Llanharry
Bridgend
↑43
Porthcawl
Peterston-super-Ely
Colwinston
Cowbridge
Llantwit Major
Parracombe
Lynmouth
Porlock
Minehead
Watchet
fracombe
Simonsbath ↓34
Wheddon Cross
44
aple

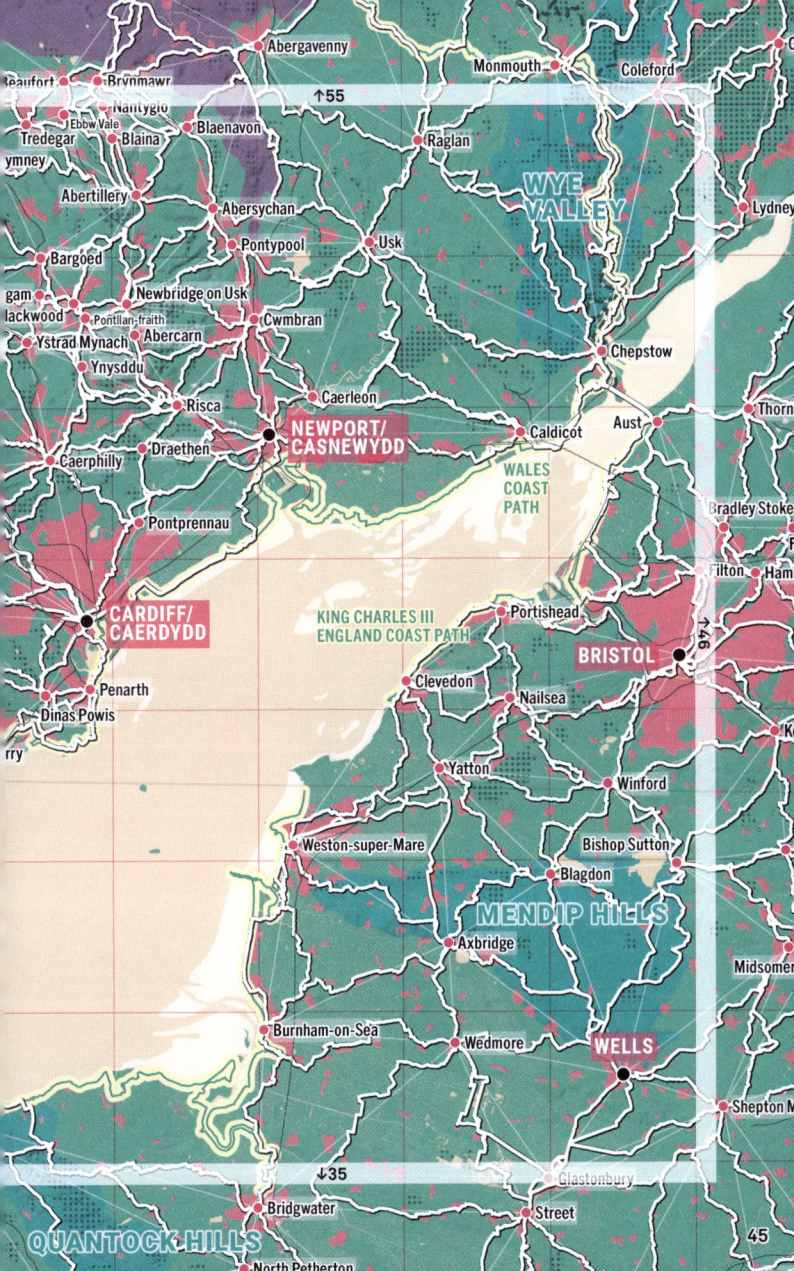

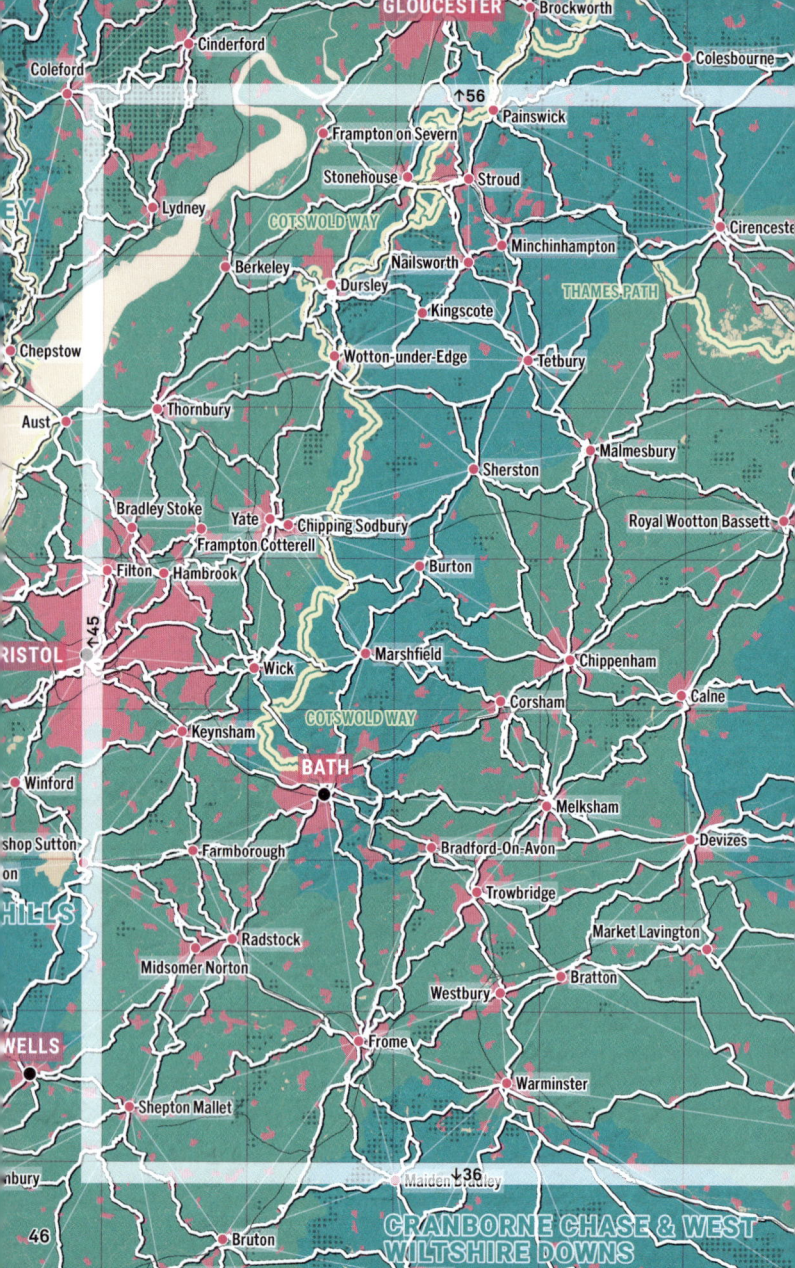

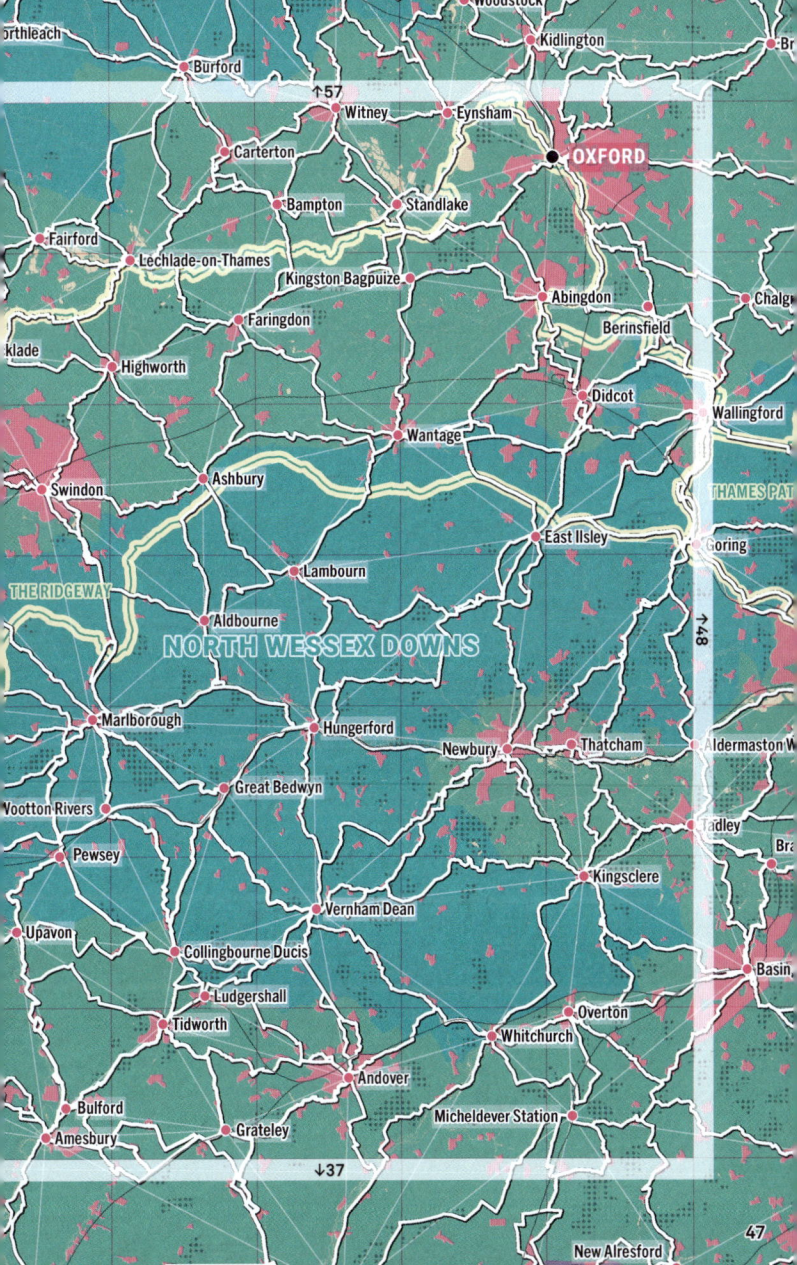

↑57

↑48

↓37

47

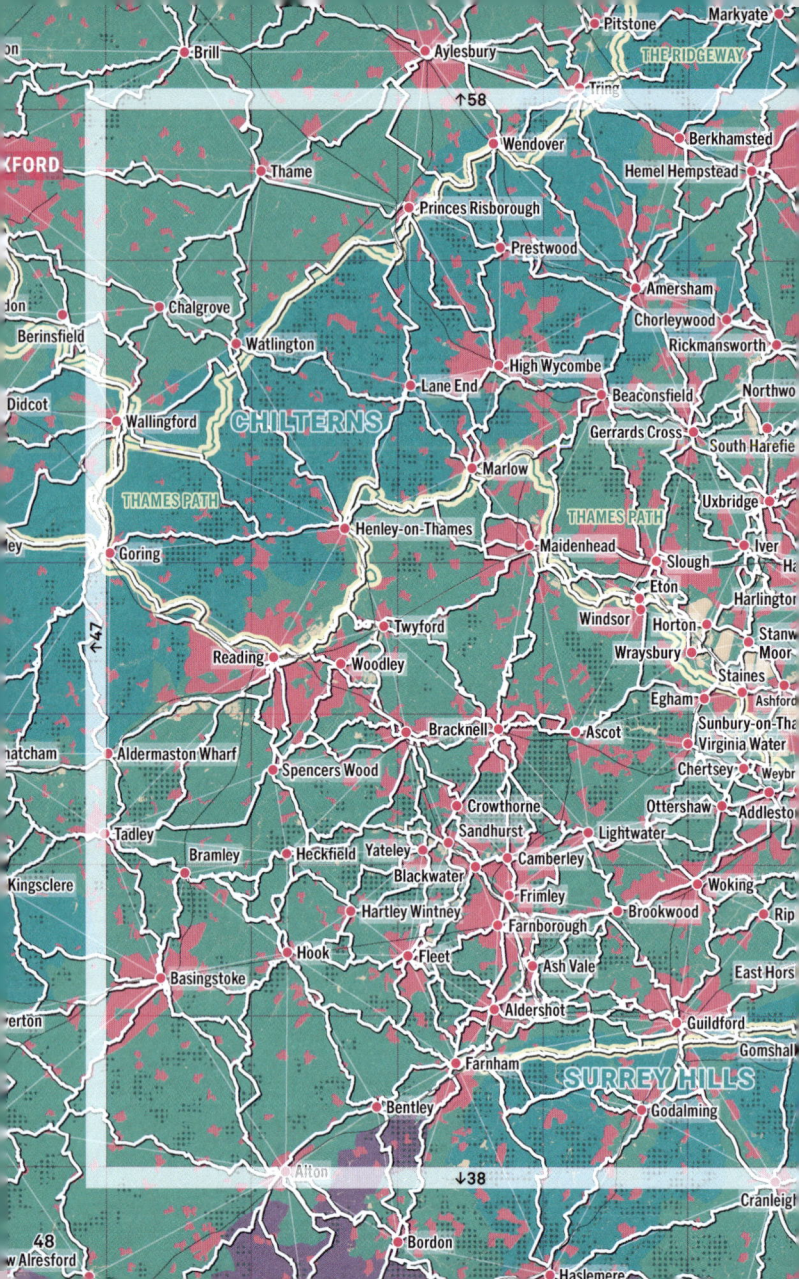

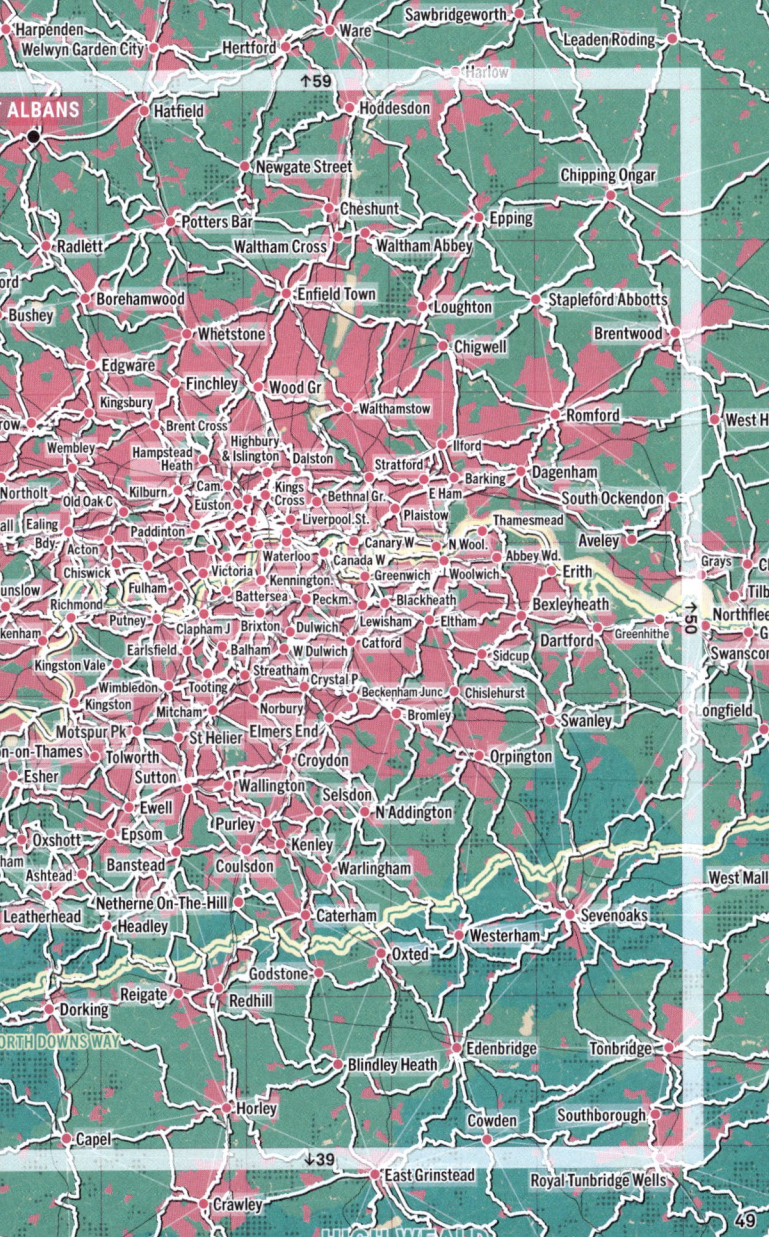

Rivers know this: there is no hurry.
We shall get there some day.
Winnie the Pooh

Rivers know this: there is no hurry.
We shall get there some day.
Winnie the Pooh

↑61

Clacton-on-Sea

Margate
Broadstairs
Herne Bay
hitstable
Minster
Ramsgate
KING CHARLES III
ENGLAND COAST PATH
ANTERBURY
Sandwich
hartham
Northbourne
Deal
Barham
KENT DOWNS
Lyminge
Hawkinge
Dover
↓41
Folkestone
Hythe

Even the smallest person can
change the course of the future.
Galadriel

PEMBROKESHIRE
COAST

PEMBRO
COAST P.

Goodwick
Fishguard

Newport

ST DAVIDS /
TYDDEWI

Newgale

Haverfordwest

Pembroke Dock

Milford Haven

Neyland

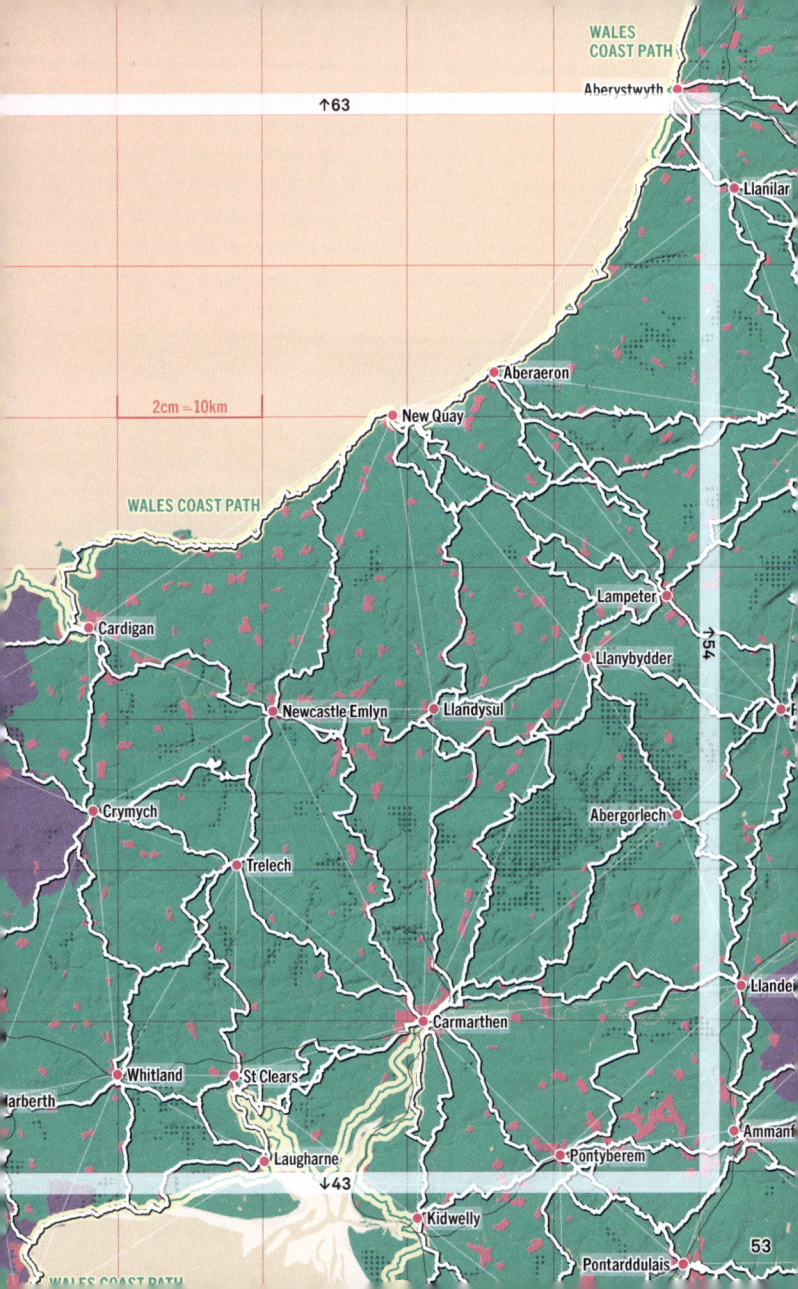

Aberystwyth

Llanilar

↑63

Aberaeron

2cm =10km

New Quay

WALES COAST PATH

Lampeter

↑54

Llanybydder

Cardigan

Newcastle Emlyn

Llandysul

Crymych

Abergorlech

Trelech

Llande

Carmarthen

Amman

Whitland

St Clears

arberth

Laugharne

Pontyberem

↓43

Kidwelly

Pontarddulais

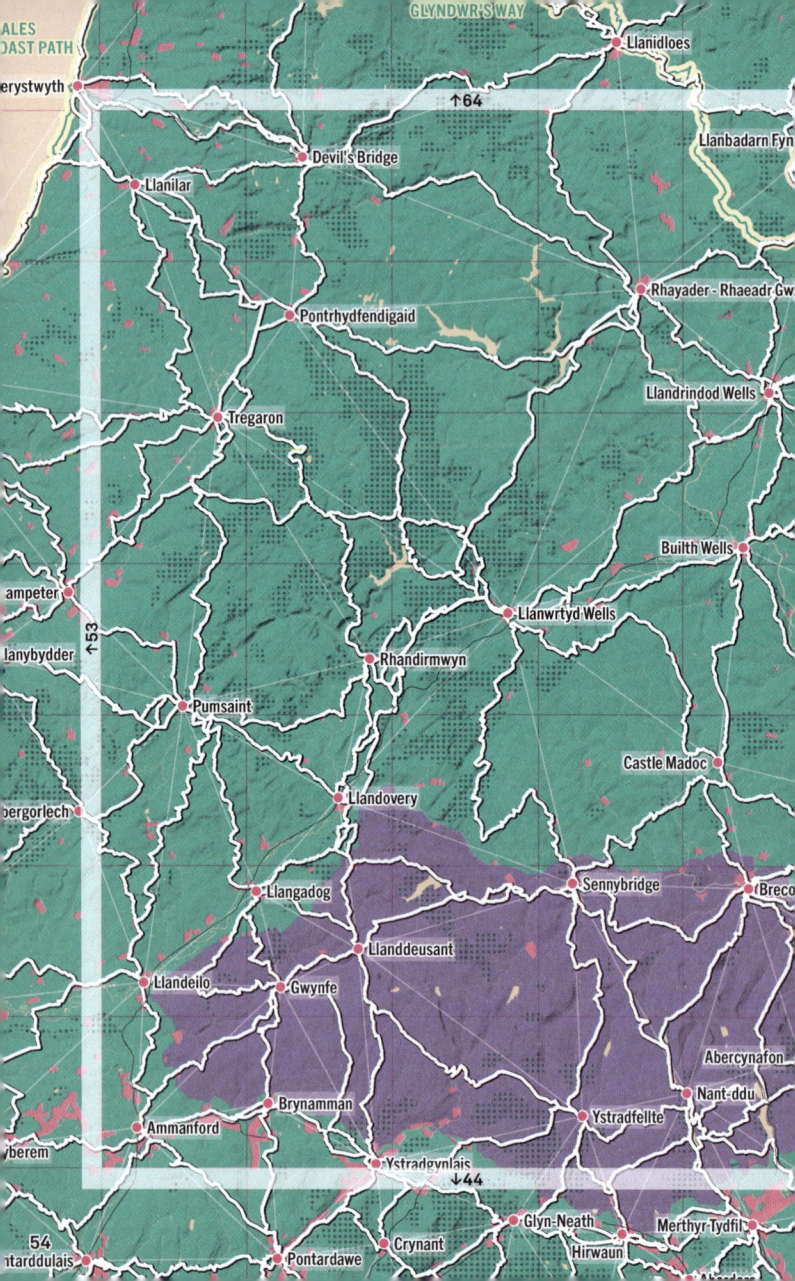

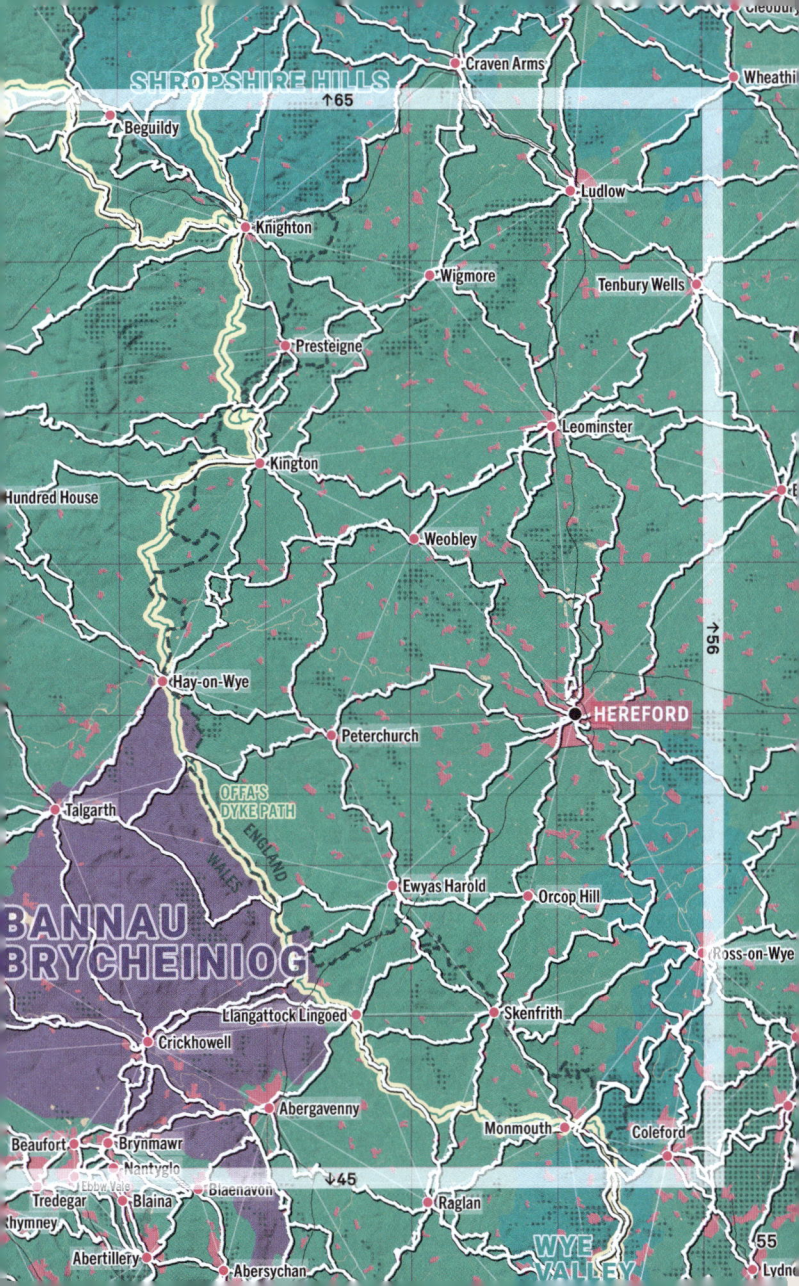

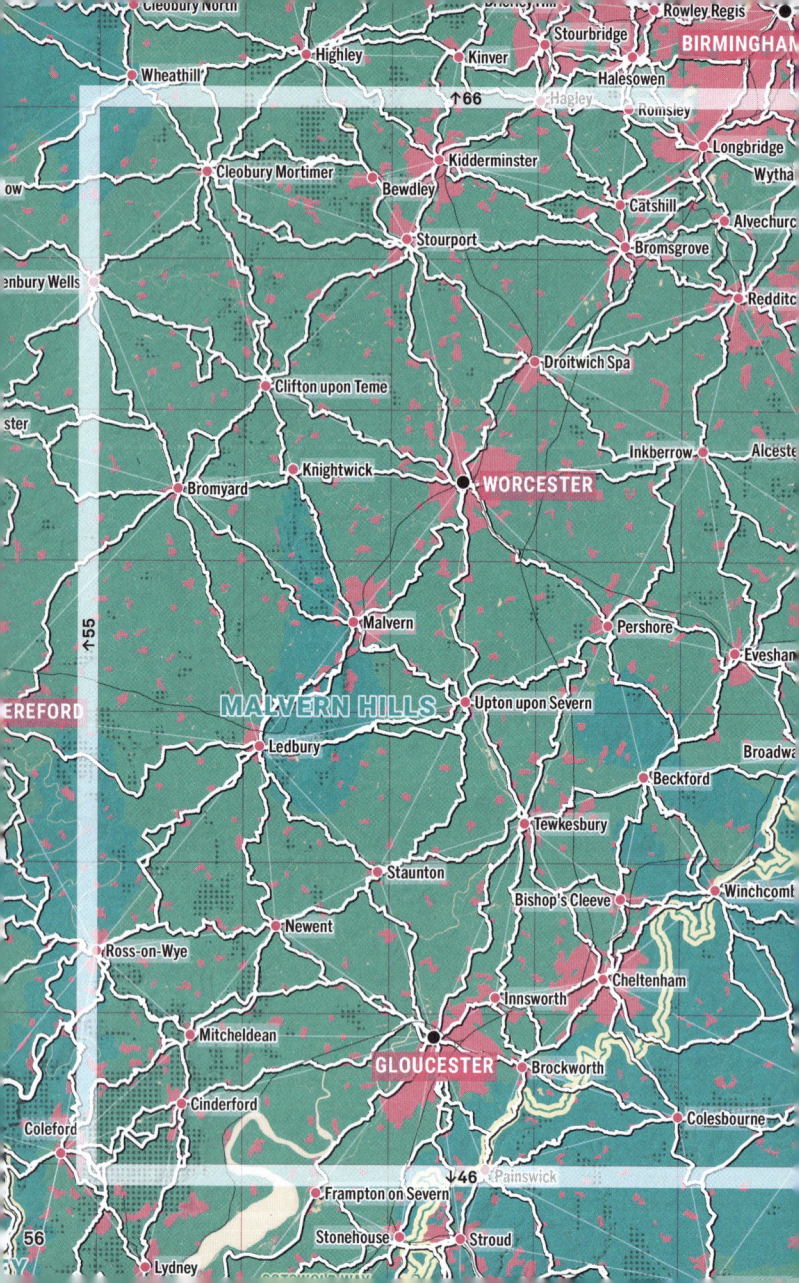

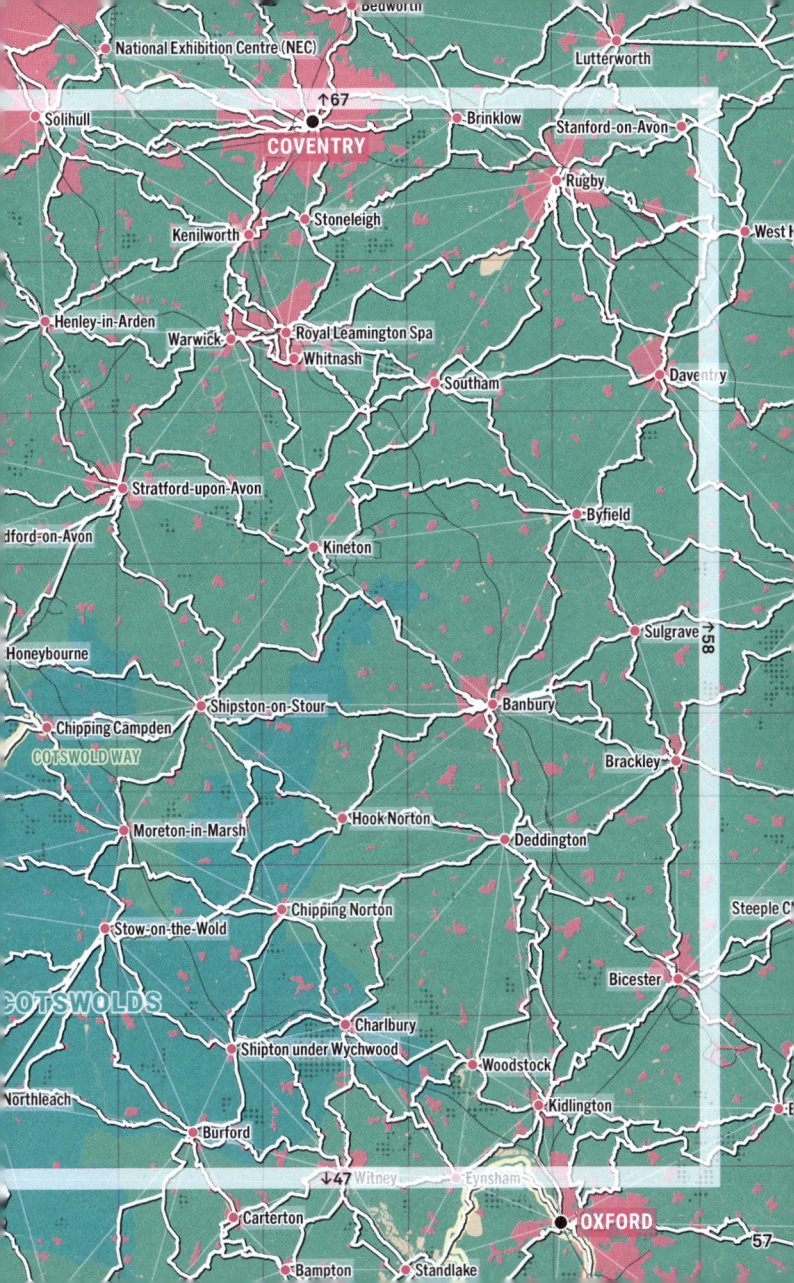

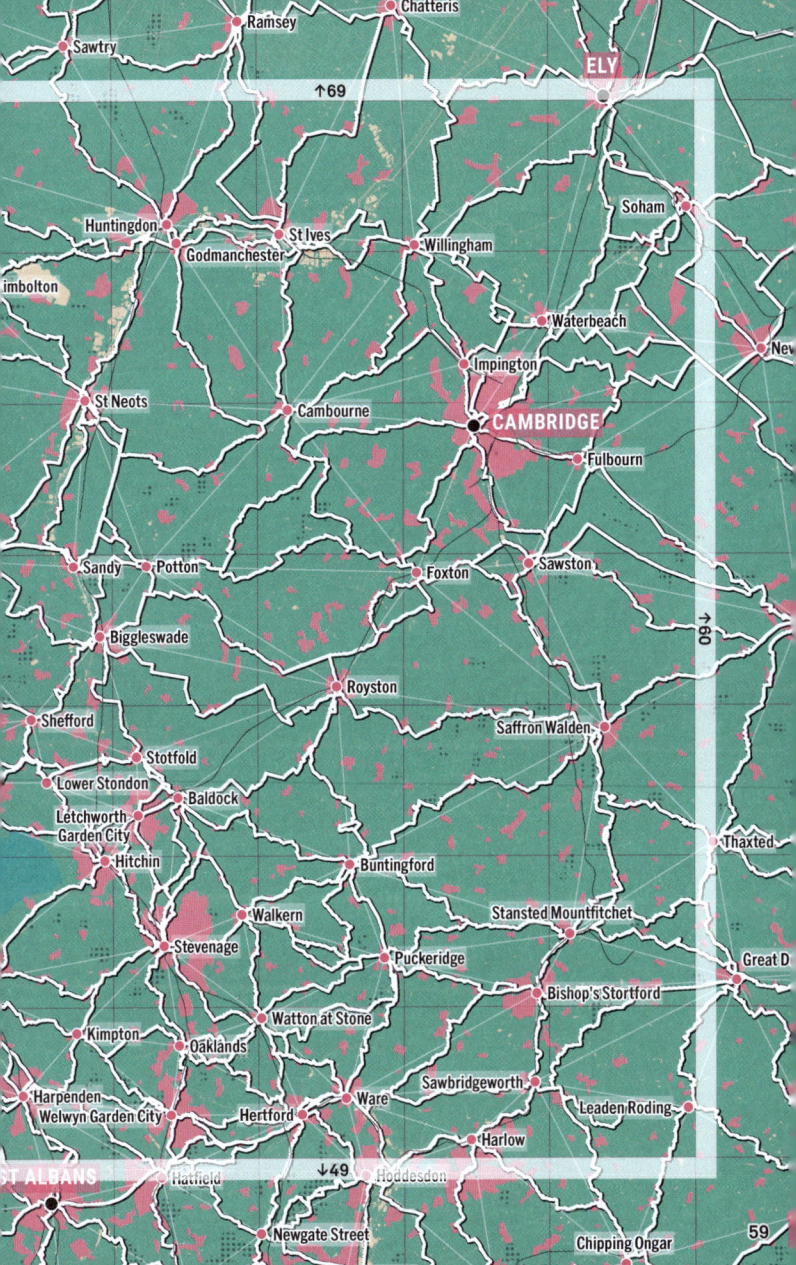

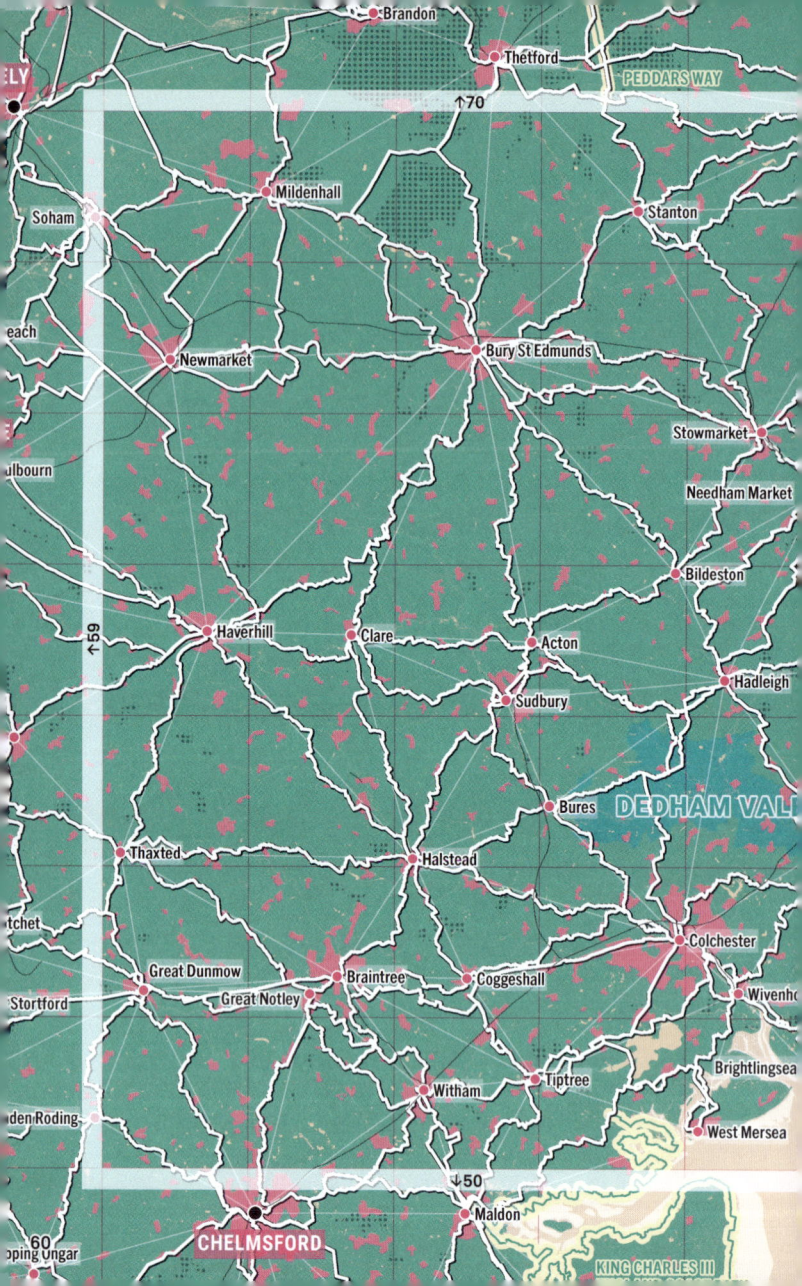

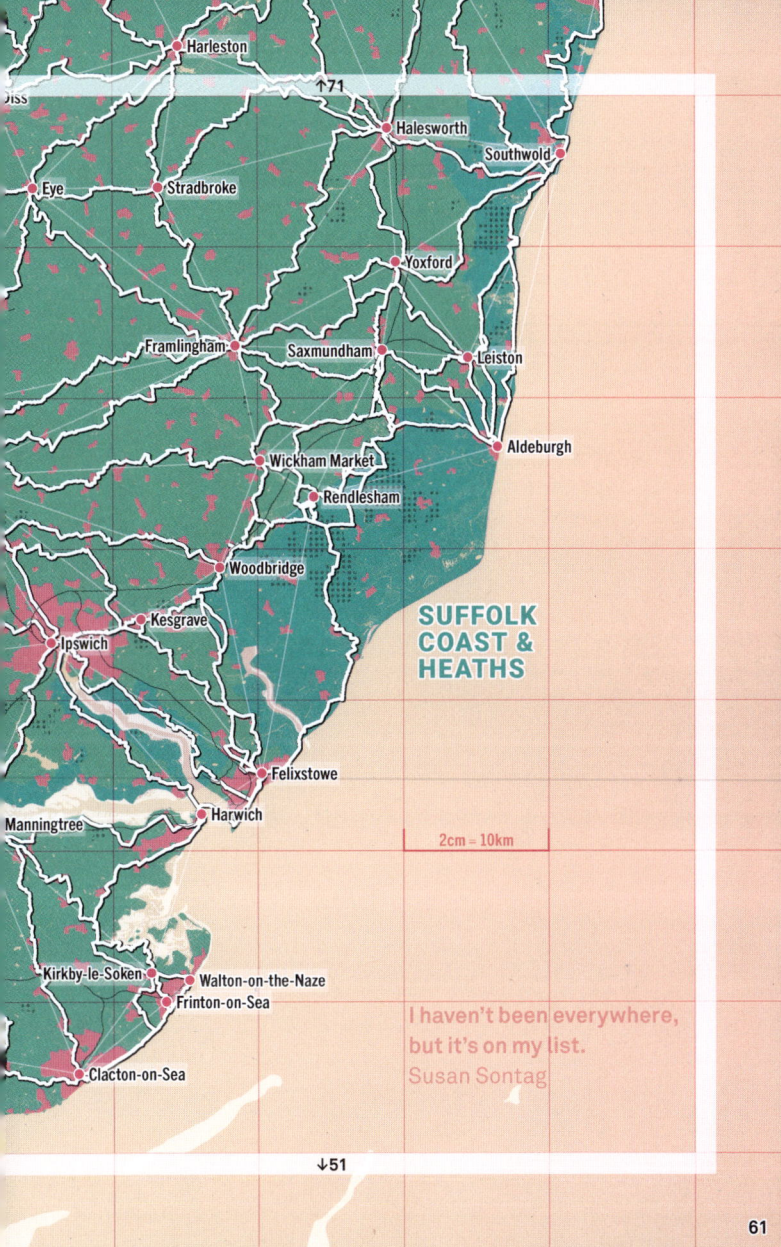

Harleston

Diss

Eye Stradbroke

Halesworth

Southwold

Yoxford

Framlingham Saxmundham

Leiston

Wickham Market Rendlesham

Aldeburgh

Woodbridge

**SUFFOLK
COAST &
HEATHS**

Kesgrave

Ipswich

Felixstowe

Manningtree

Harwich

2cm = 10km

Kirkby-le-Soken Walton-on-the-Naze
Frinton-on-Sea

I haven't been everywhere,
but it's on my list.
Susan Sontag

Clacton-on-Sea

Roads? Where we're going,
we don't need roads.
Dr Emmett Brown

Rhosgadfan

Beddgelert

Llanaelhaearn

Nefyn

Criccieth

Porthmadog

Pwllheli

LLŶN

Harlech

Abersoch

WALES COAST PATH

↑64 Barmouth

Tywyn

You should spend 20 minutes a day out
in nature, unless you're too busy. In
that case, you should spend an hour.
Zen saying

**WALES
COAST PATH**

Aberystwyth

L63 lar

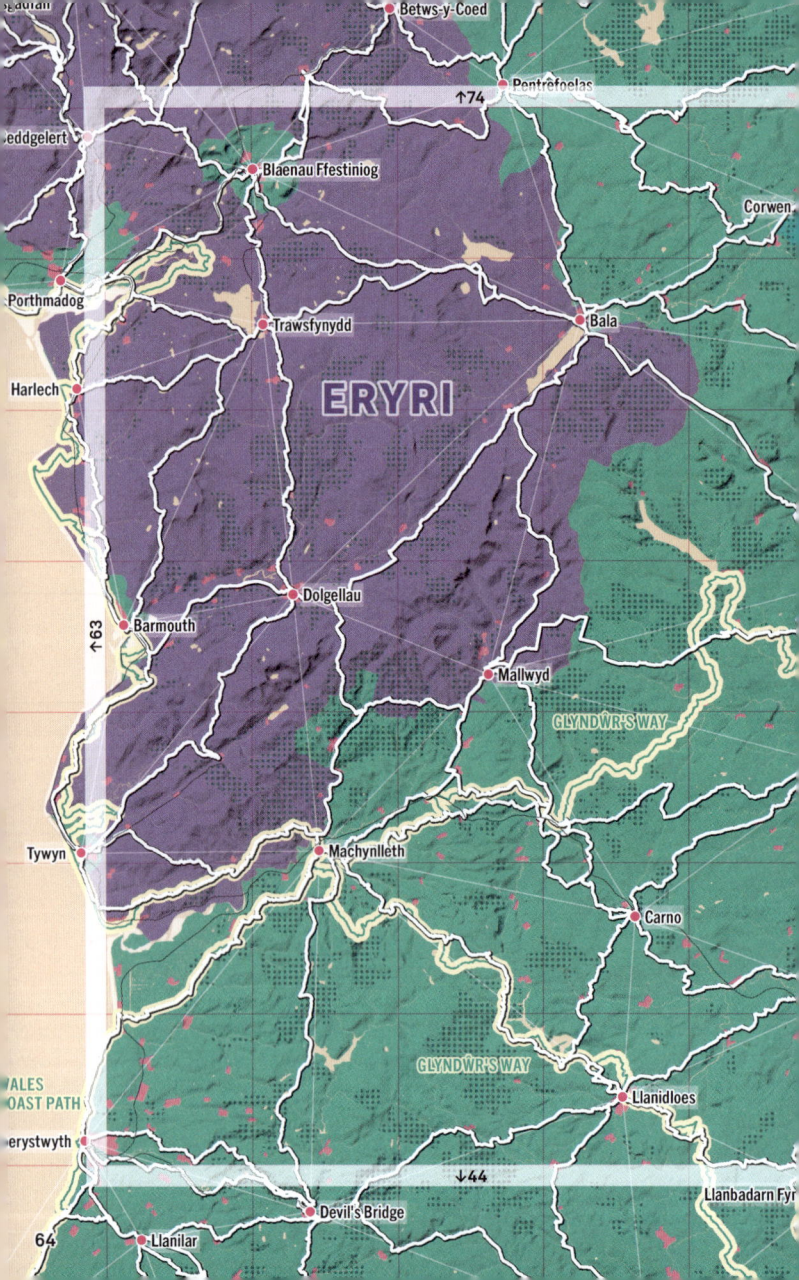

Betws-y-Coed

Pentrefoelas

↑74

eddgelert

Blaenau Ffestiniog

Corwen

Porthmadog

Trawsfynydd

Bala

Harlech

ERYRI

Dolgellau

↑63

Barmouth

Mallwyd

GLYNDWR'S WAY

Tywyn

Machynlleth

Carno

GLYNDWR'S WAY

Llanidloes

WALES
COAST PATH

erystwyth

Llanbadarn Fyr

↓44

Devil's Bridge

64   Llanilar

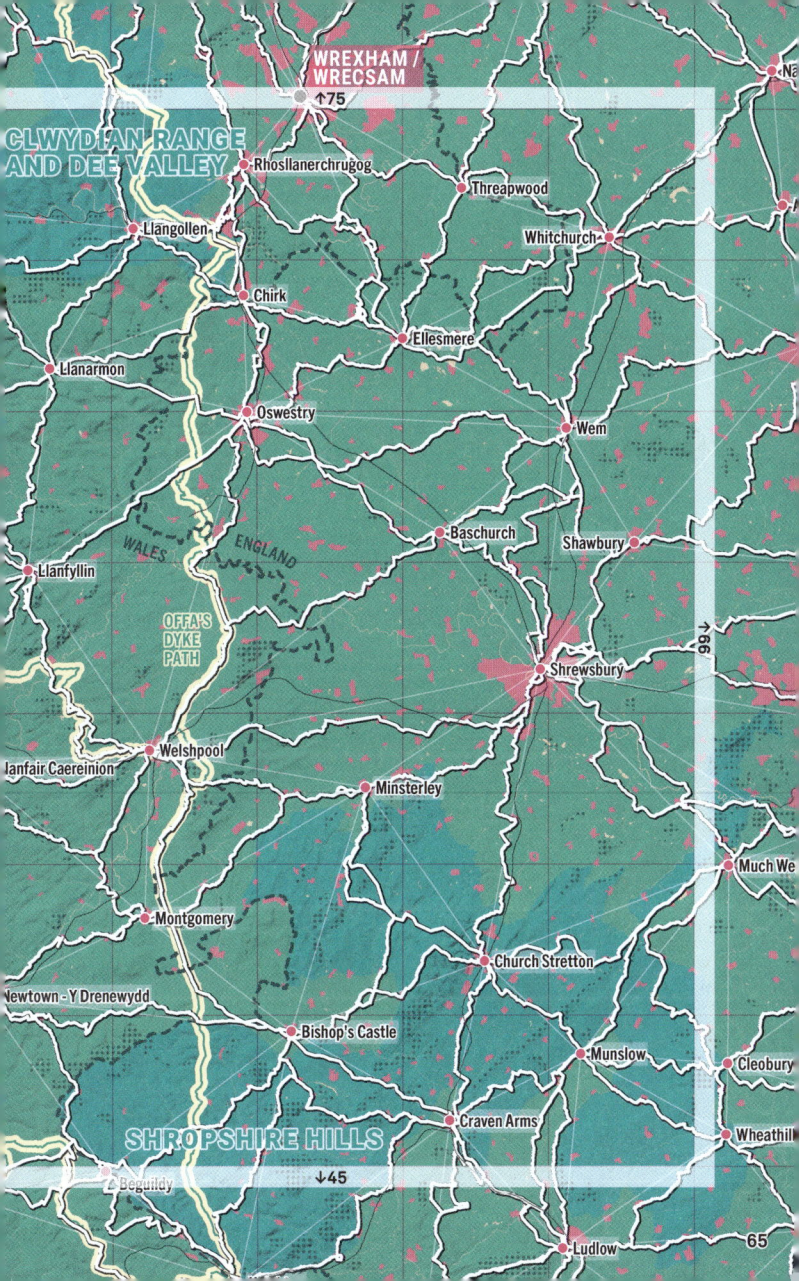

CLWYDIAN RANGE
AND DEE VALLEY

Rhosllanerchrugog

Threapwood

Na

Llangollen

Whitchurch

Chirk

Ellesmere

Llanarmon

Oswestry

Wem

WALES

ENGLAND

Baschurch

Shawbury

Llanfyllin

OFFA'S
DYKE
PATH

Shrewsbury

↑A49

Welshpool

Llanfair Caereinion

Minsterley

Much We

Montgomery

Church Stretton

Newtown - Y Drenewydd

Bishop's Castle

Munslow

Cleobury

Craven Arms

Wheathil

SHROPSHIRE HILLS

↓45

Beguildy

Ludlow

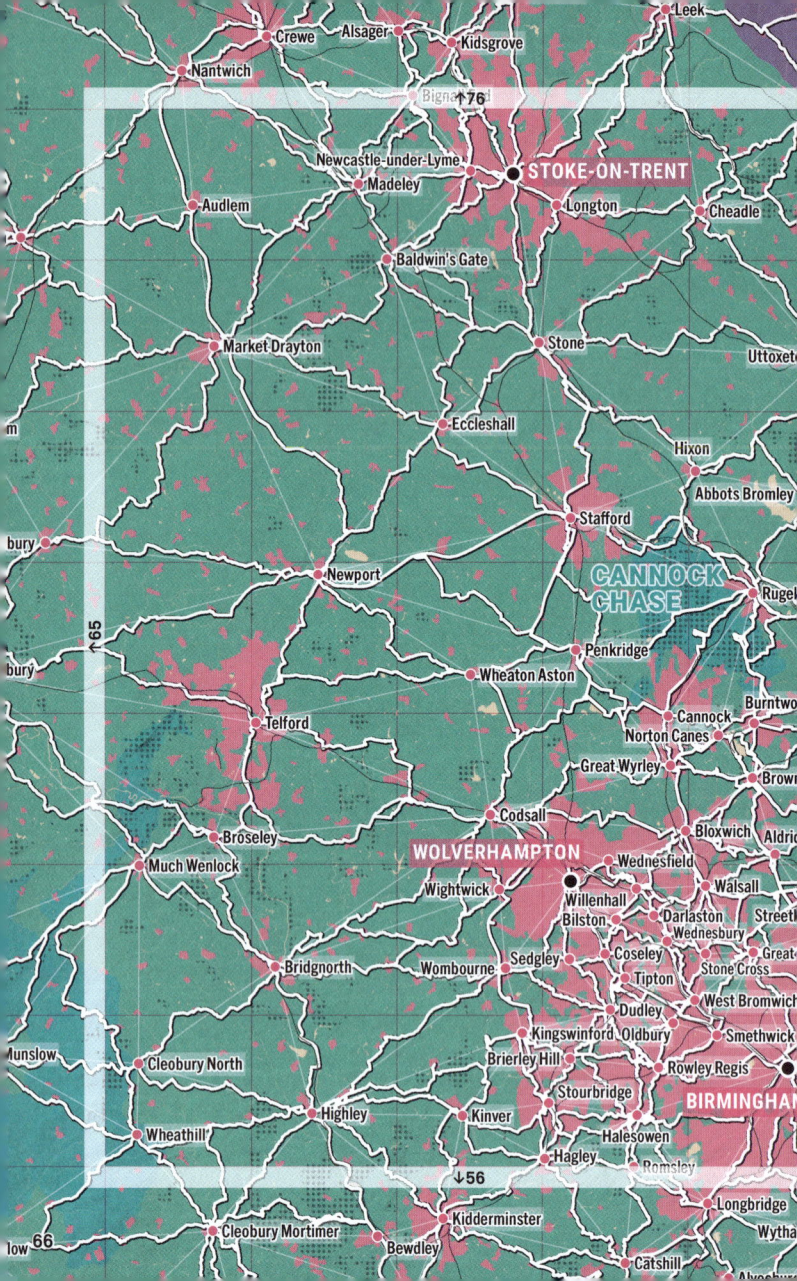

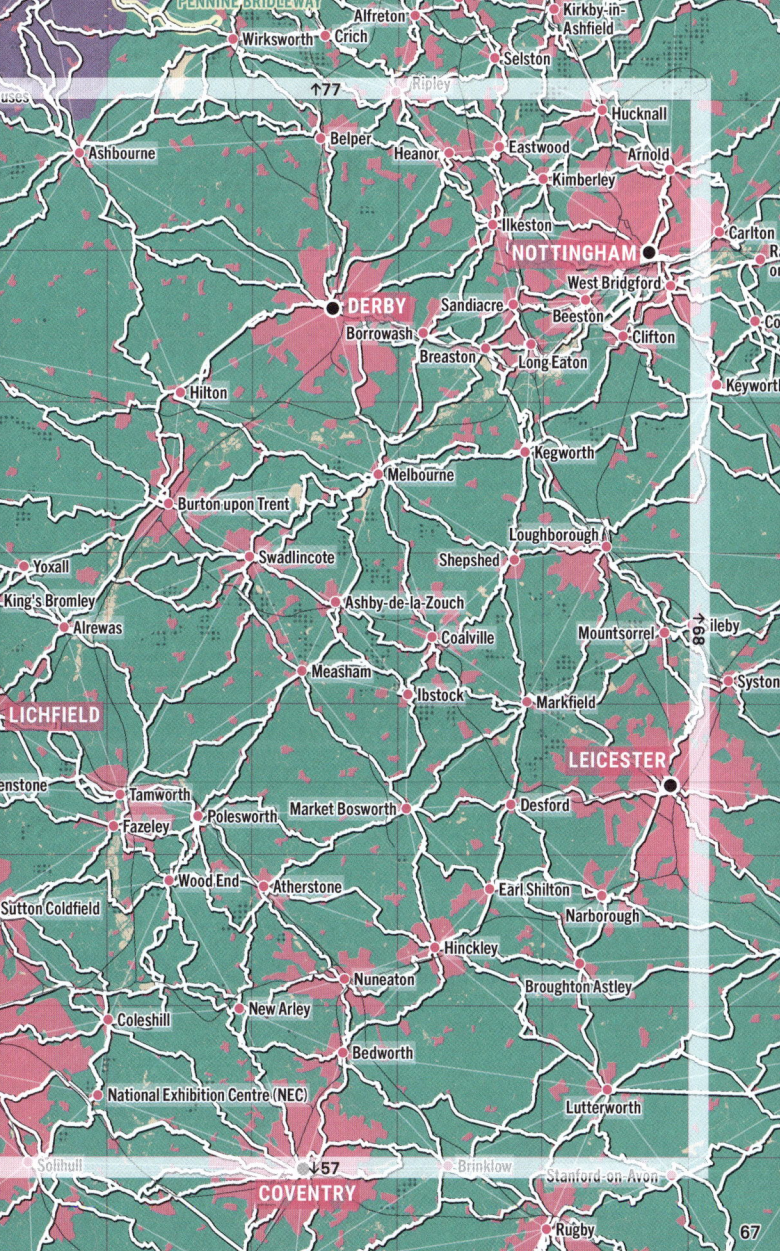

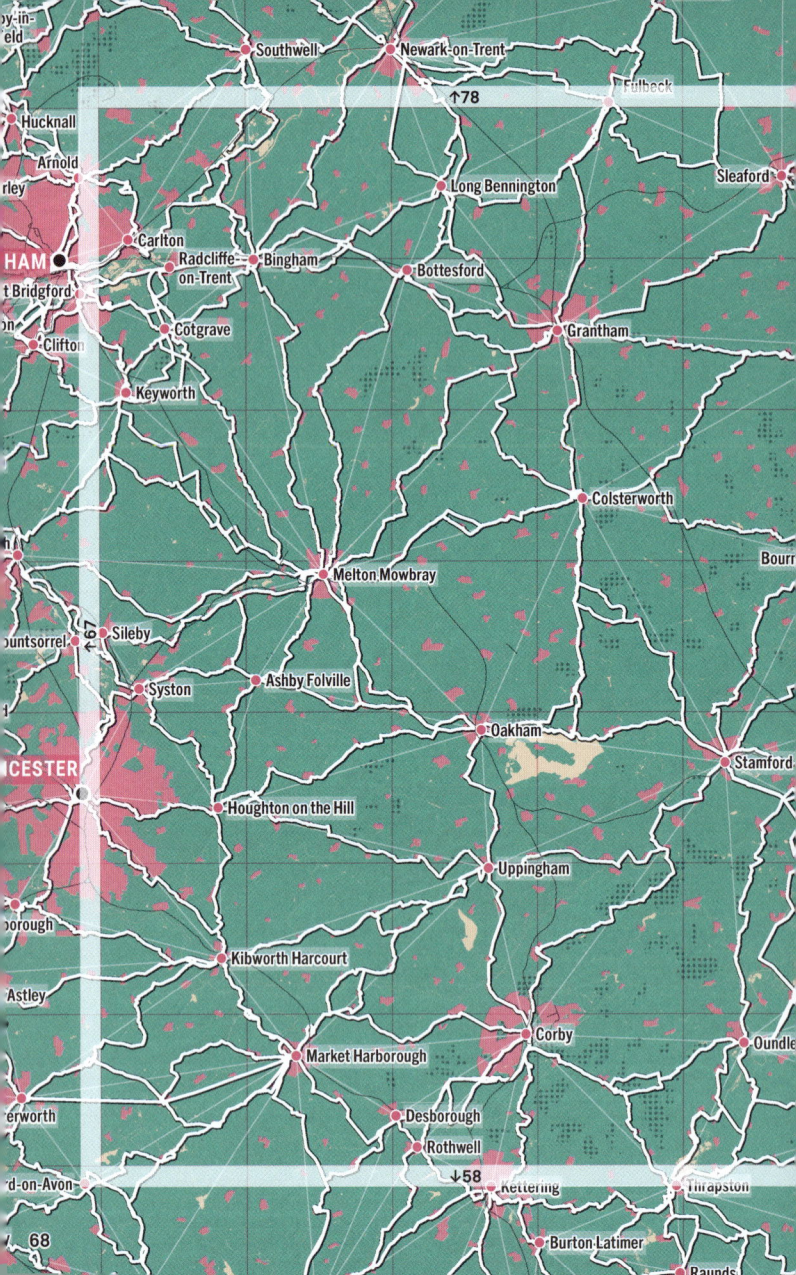

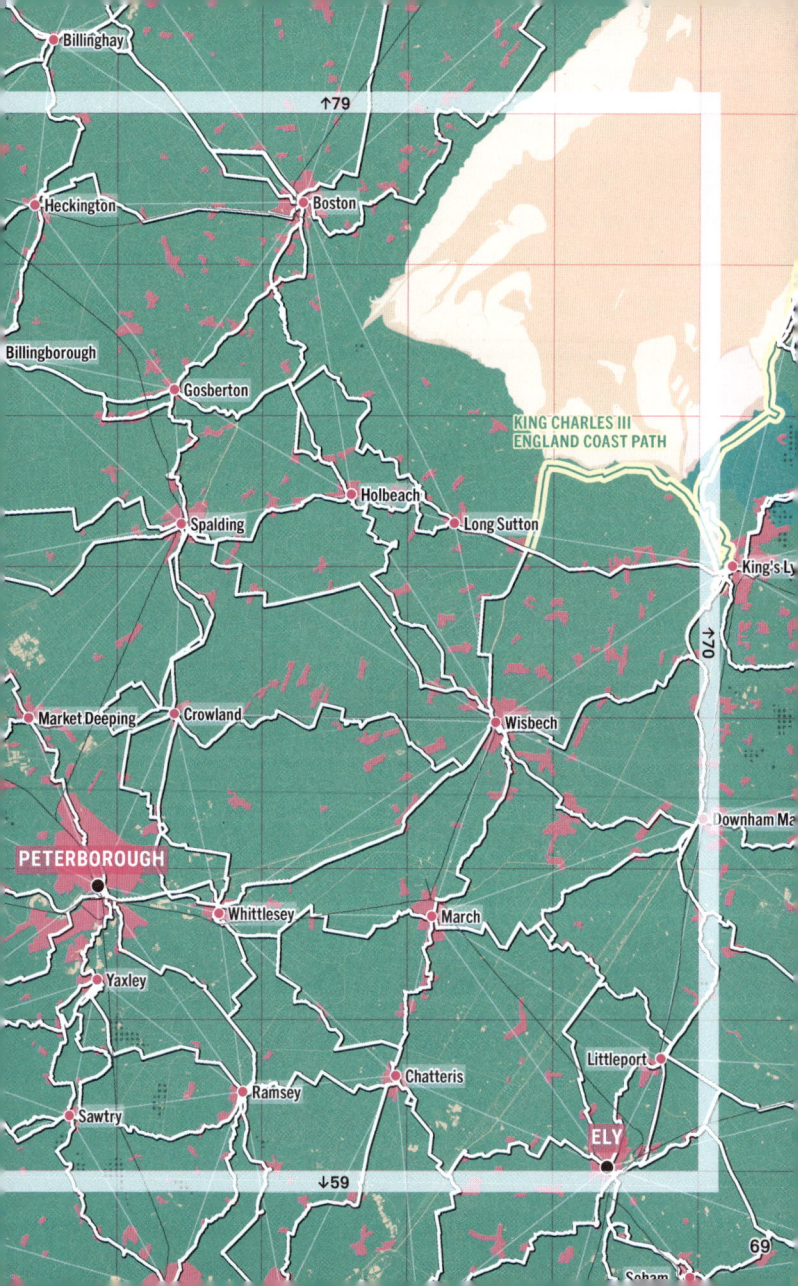

Billinghay

Heckington

Boston

Billingborough

Gosberton

KING CHARLES III
ENGLAND COAST PATH

Holbeach

Spalding

Long Sutton

King's Ly

Market Deeping

Crowland

Wisbech

Downham Ma

PETERBOROUGH

Whittlesey

March

Yaxley

Littleport

Sawtry

Ramsey

Chatteris

ELY

Soham

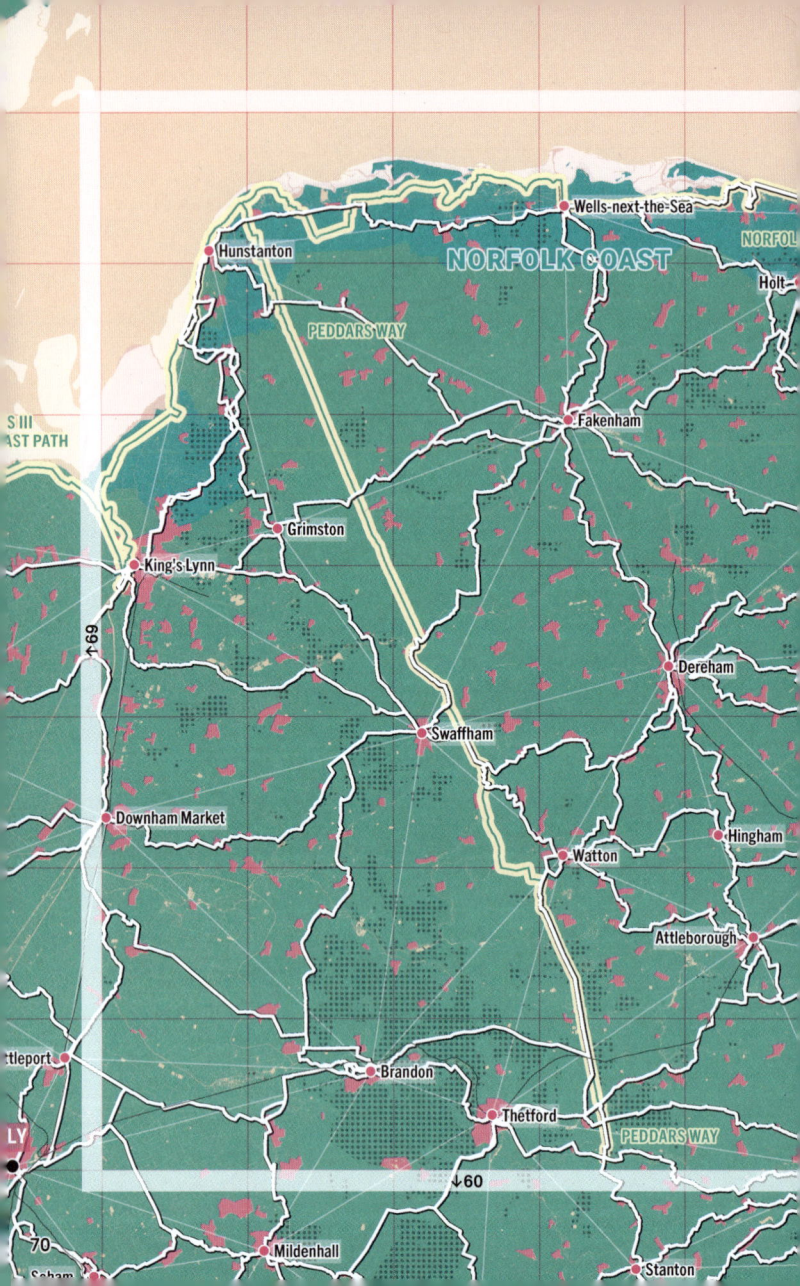

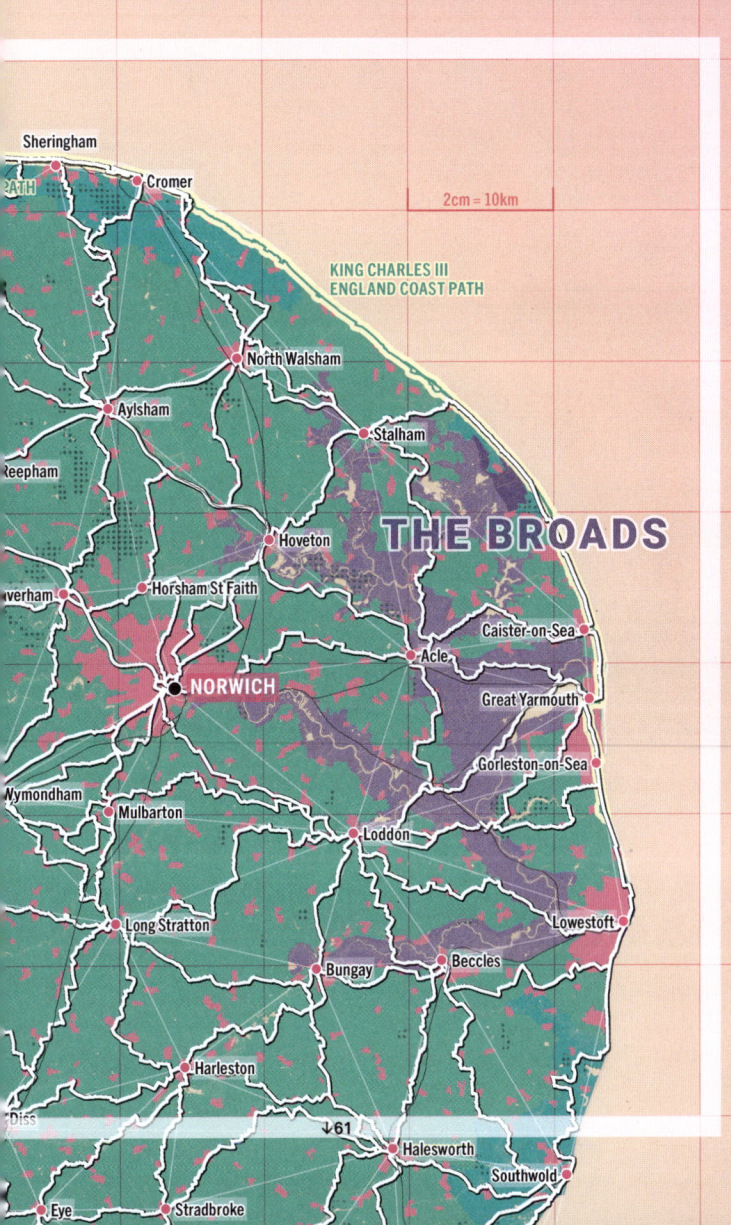

2cm = 10km

KING CHARLES III
ENGLAND COAST PATH

THE BROADS

Sheringham

Cromer

PATH

North Walsham

Aylsham

Reepham

Stalham

Hoveton

Horsham St Faith

verham

Caister-on-Sea

Acle

NORWICH

Great Yarmouth

Gorleston-on-Sea

Wymondham

Mulbarton

Loddon

Long Stratton

Lowestoft

Bungay

Beccles

Harleston

Diss

↓61

Halesworth

Southwold

Eye

Stradbroke

You must be imaginative, strong-hearted. You must try things that may not work, and you must not let anyone define your limits because of where you come from. Gusteau

↓62

The function of freedom is
to free someone else.
Toni Morrison

2cm = 10km

WALES
COAST
PATH

Holyhead

ANGLESEY

↑74

Llangefni

Beaumaris

Menai Bridge

BANGOR

Bethesd.

Caernarfon

Llanberis

Rhosgadfan

↓63

Beddgelert

Llanaelhaearn

73

↑80

Life moves pretty fast. If you don't stop and look around once in a while, you could miss it.
Ferris Bueller

2cm = 10km

↑73

Prestatyn

Llandudno

Kinmel Bay

Rhyl

Colwyn Bay

Conwy

Abergele

Rhuddlan

Llandudno Junction

Penmaenmawr

Beaumaris

Llanfairfechan

ST ASAPH / LLANELWY

ridge

BANGOR

Llanfair Talhaiarn

Bethesda

Denbigh

anberis

Llanrwst

Gwytherin

gadfan

Betws-y-Coed

eddgelert

Pentrefoelas

↓64

74

Blaenau Ffestiniog

Corwen

ENGLAND COAST PATH

↑81

Bamber Bridge
Tarleton
Leyland
Wheelton
White Copp
Southport
Eccleston
Chorley
Coppull
Adlington
Horwic
Burscough
Parbold
Standish
Blackrod
Aspull
Lostoc
We
Ormskirk
Skelmersdale
Wigan
Hindley
Formby
Downholland
Orrell
Ince-in-Makerfield
Platt Bridge
Stanley Gate
Maghull
Rainford
Billinge
L
Crosby
Kirkby
Ashton-in-Makerfield
Golborne
Litherland
Eccleston
Haydock
Newton-le-
Bootle
St Helens
Willows
Walton
Prescot
Lea
Burtonwood
Wallasey
Green
LIVERPOOL
Hoylake
Birkenhead
Belle Vale
Warrington
West Kirby
Woodchurch
Widnes
↓76
Bebington
Garston
Runcorn
Stretton
Heswall
Neston
Frodsham
Holywell
Ellesmere Port
Helsby
Sandiway
Caerwys
Flint
Wi
ENGLAND
Shotton
Kelsall
CHESTER
Mold
Buckley
TH
Tarporley
Ruthin
WALES
Tattenhall
WREXHAM /
WRECSAM
↓65
CLWYDIAN RANGE
AND DEE VALLEY
Rhosllanerchrugog
Threapwood

WALES
COAST PATH

75

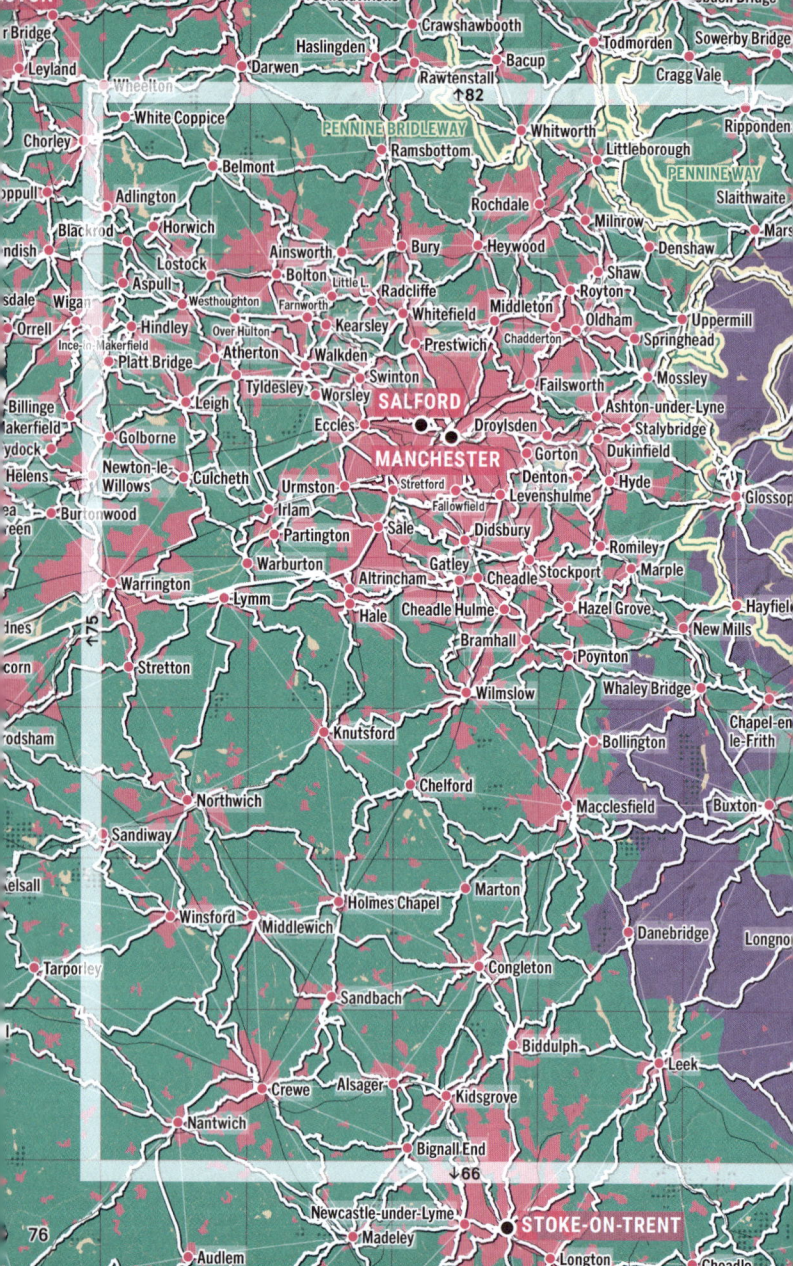

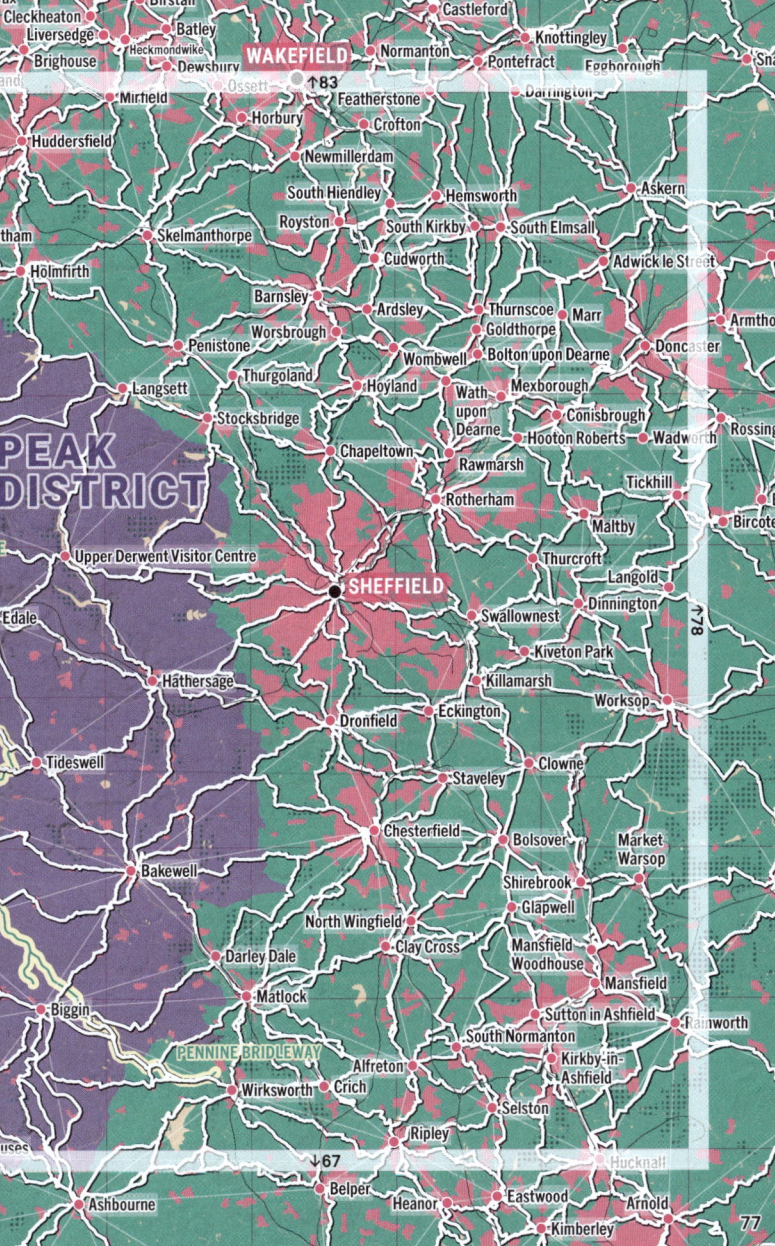

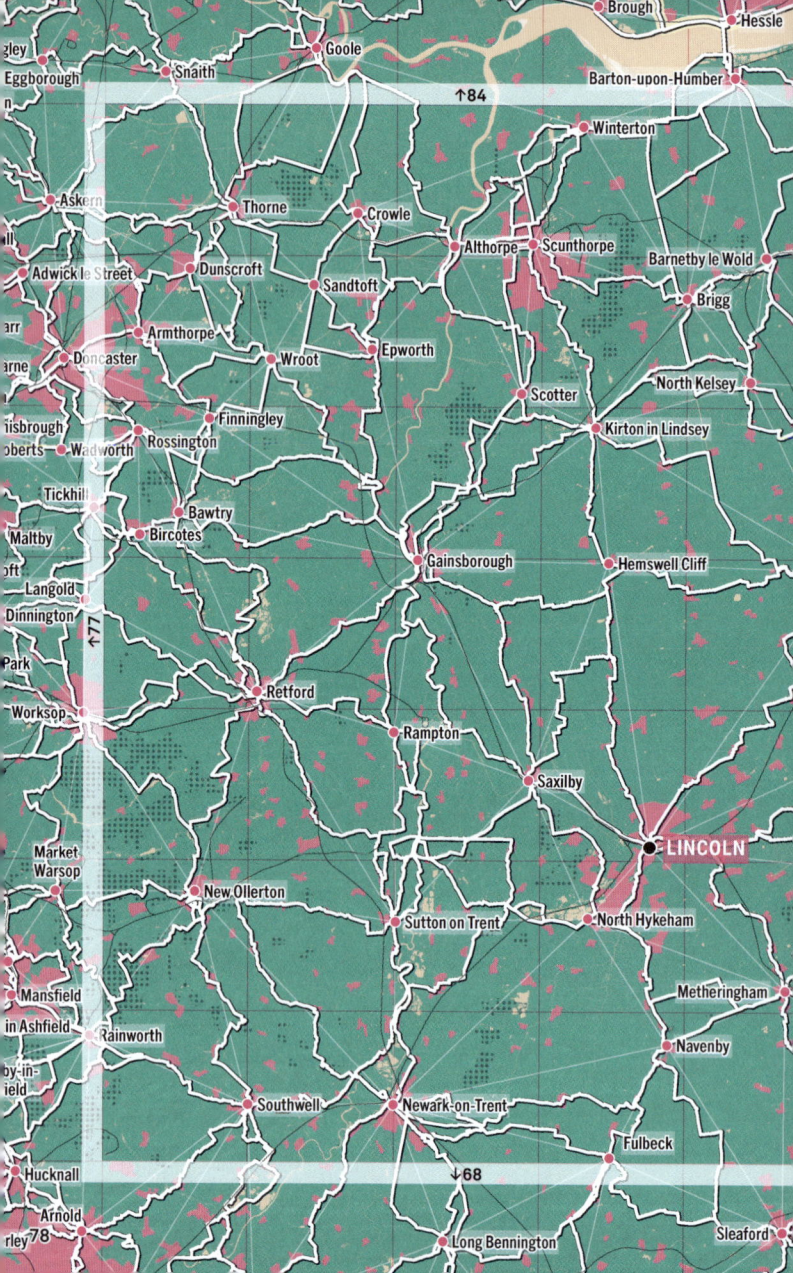

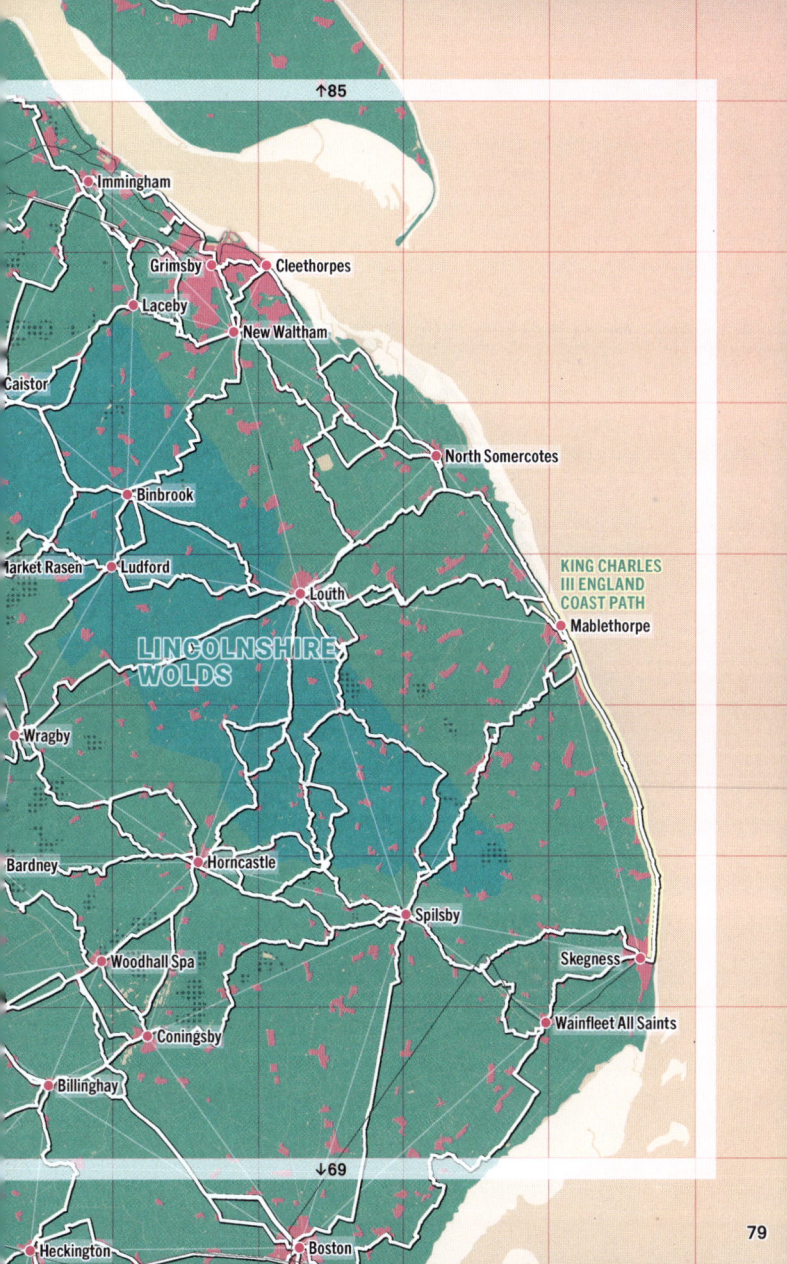

Immingham

Grimsby · Cleethorpes

Laceby

New Waltham

Caistor

North Somercotes

Binbrook

Market Rasen · Ludford

KING CHARLES
III ENGLAND
COAST PATH

Louth

Mablethorpe

LINCOLNSHIRE
WOLDS

Wragby

Bardney

Horncastle

Spilsby

Woodhall Spa

Skegness

Coningsby

Wainfleet All Saints

Billinghay

Heckington

Boston

↑88

Anytime I feel lost, I
pull out a map and
stare. I stare until
I have reminded
myself that life is a
giant adventure, so
much to do, to see.
Angelina Jolie

↓74

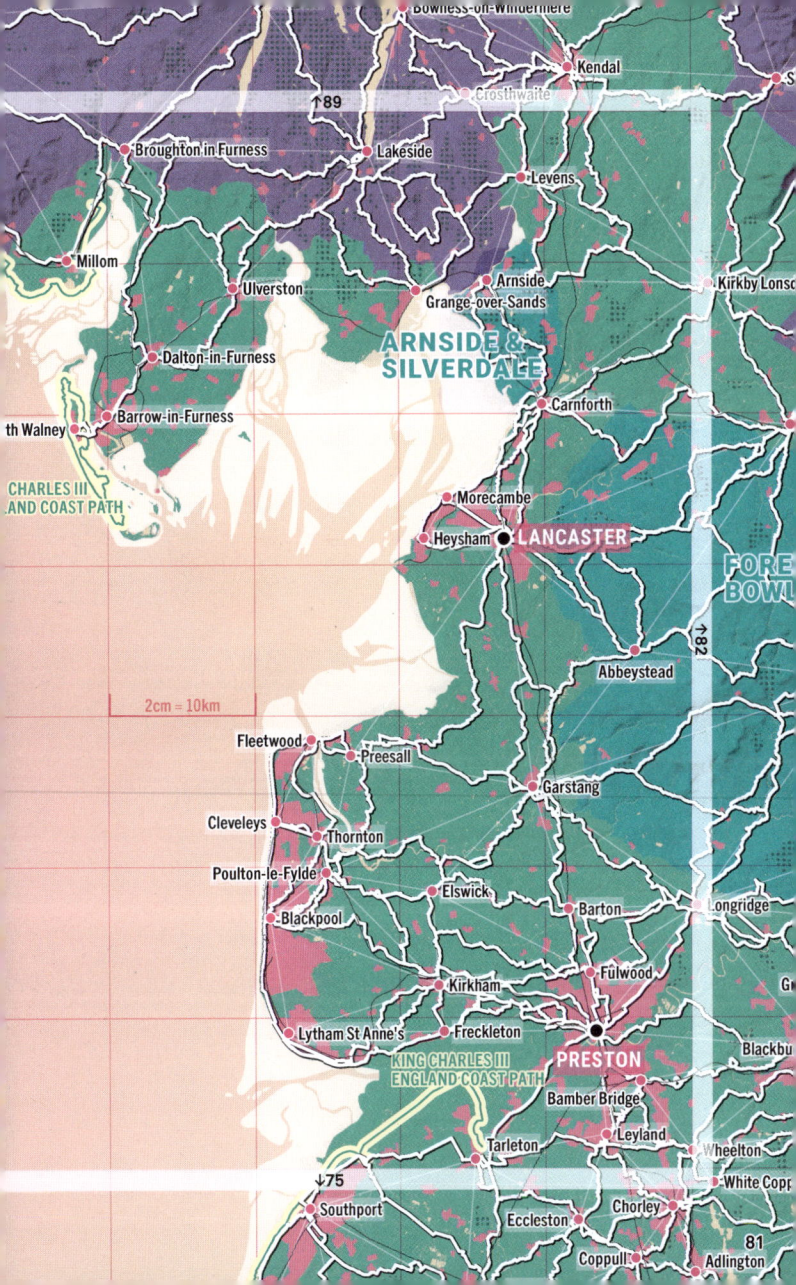

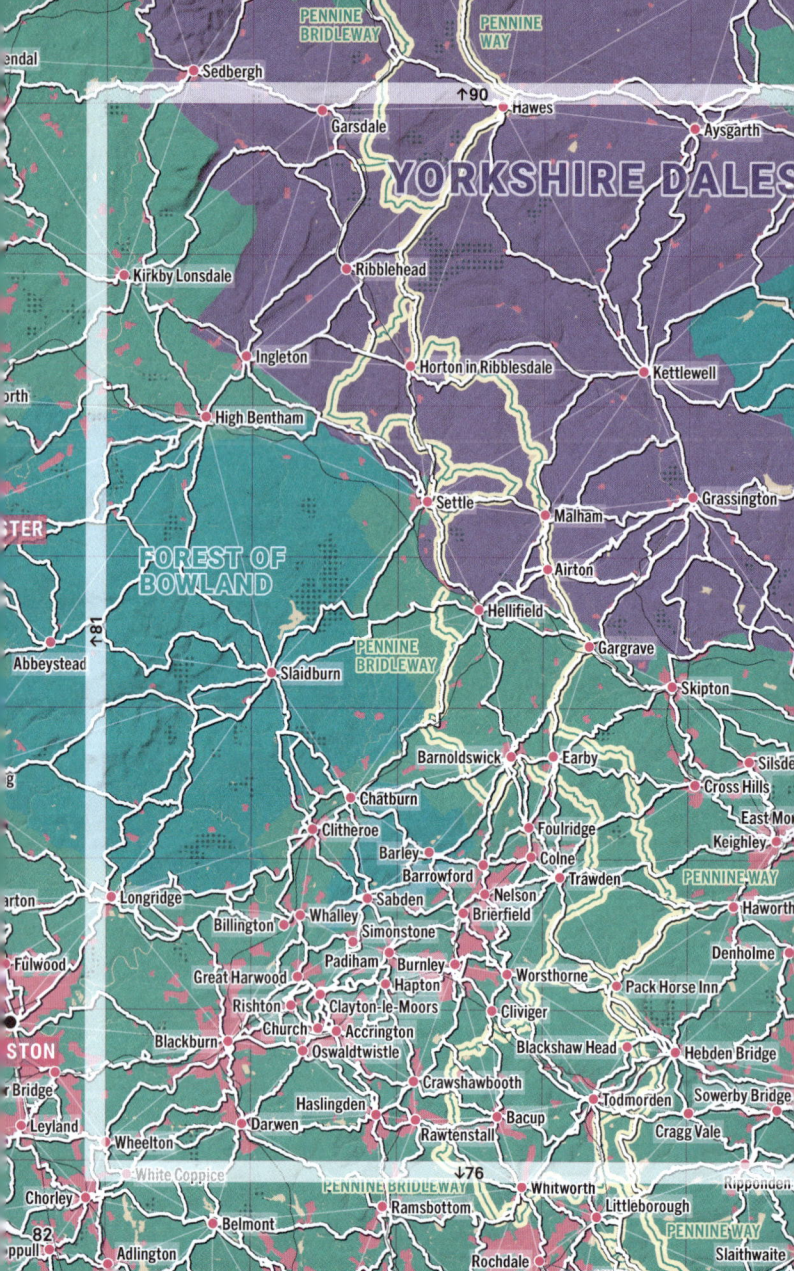

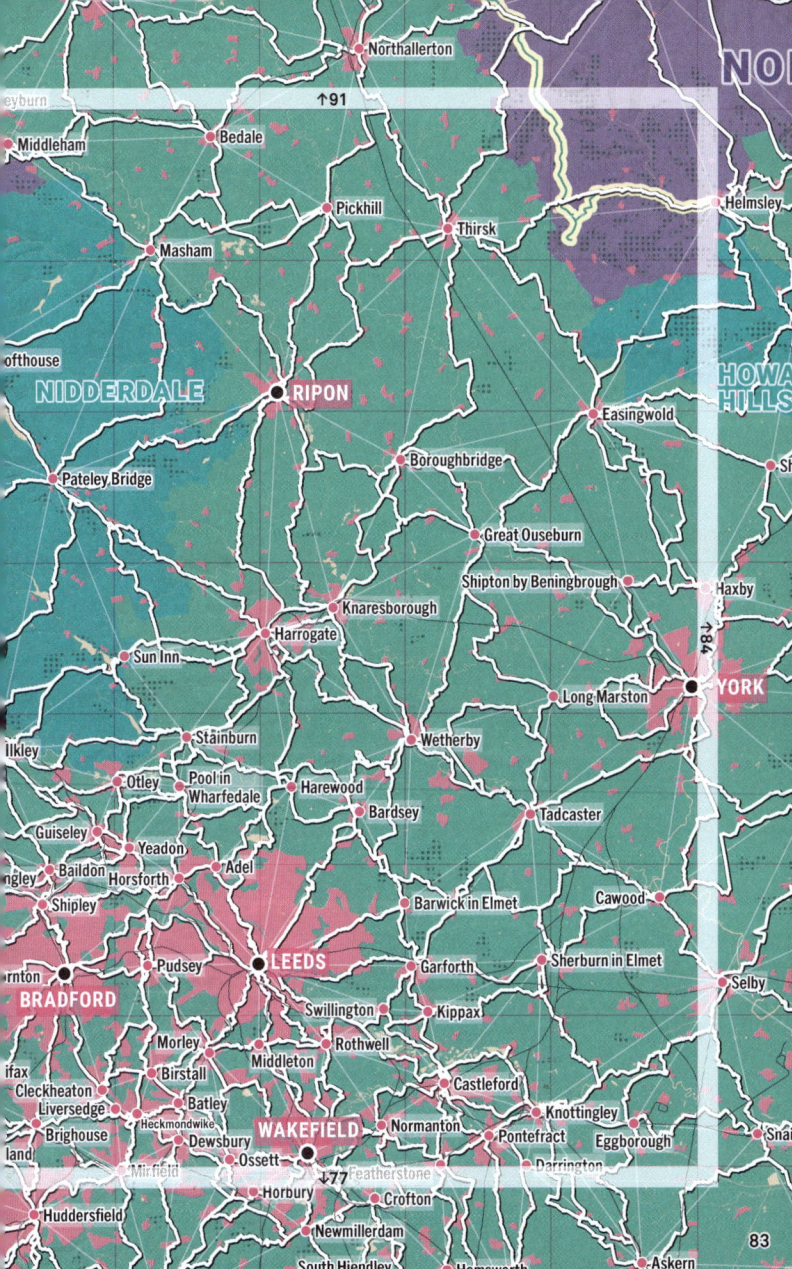

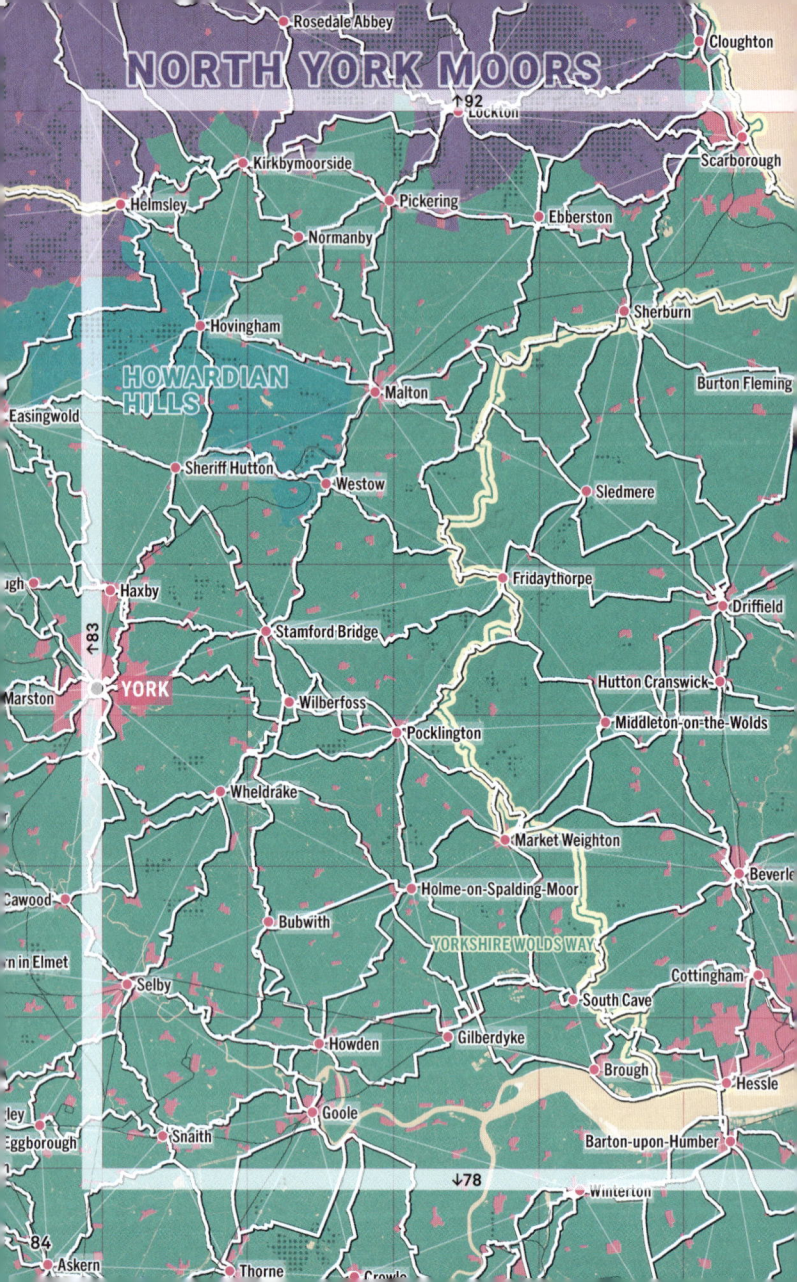

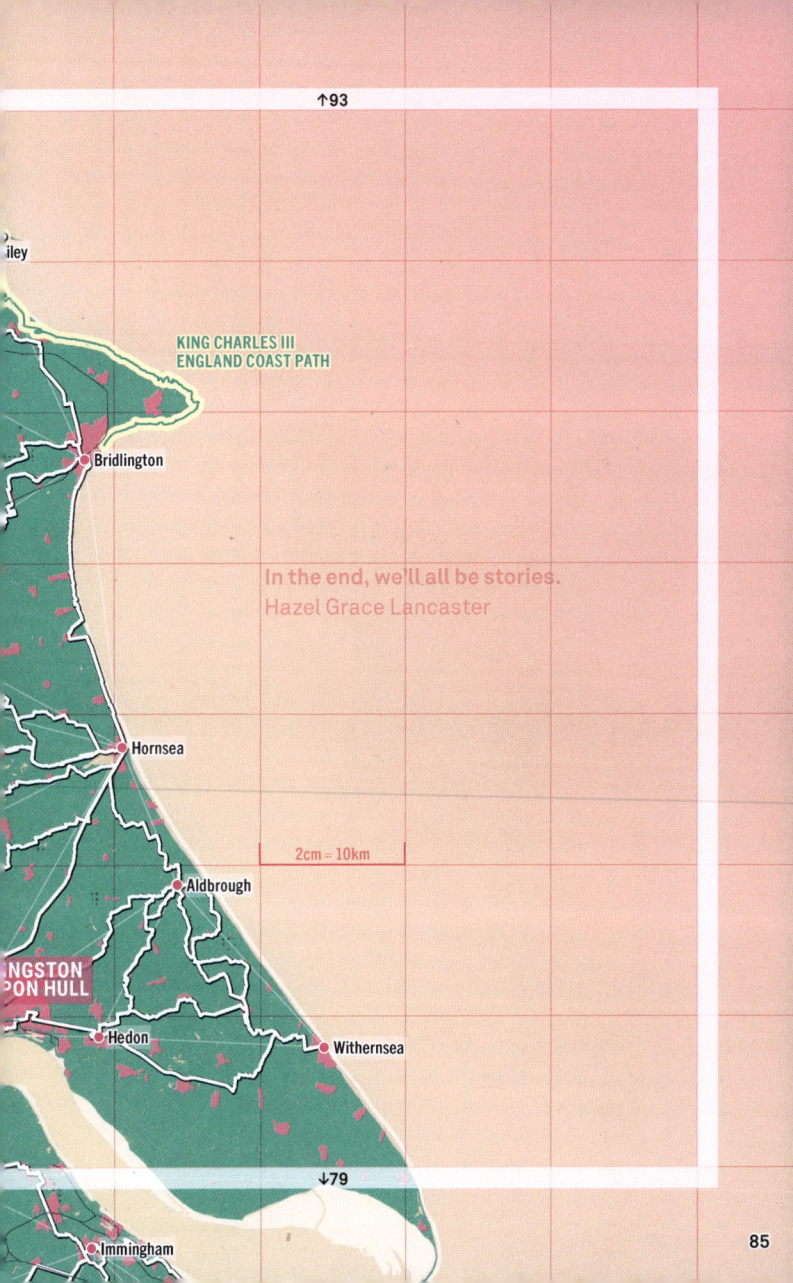

iley

**KING CHARLES III
ENGLAND COAST PATH**

● Bridlington

In the end, we'll all be stories.
Hazel Grace Lancaster

● Hornsea

2cm = 10km

● Aldbrough

NGSTON
PON HULL

● Hedon   ● Withernsea

● Immingham

SOUTHERN UPLAND WAY • Portpatrick

Stranraer

All we have to decide is
what to do with the time
that is given to us.
Gandalf

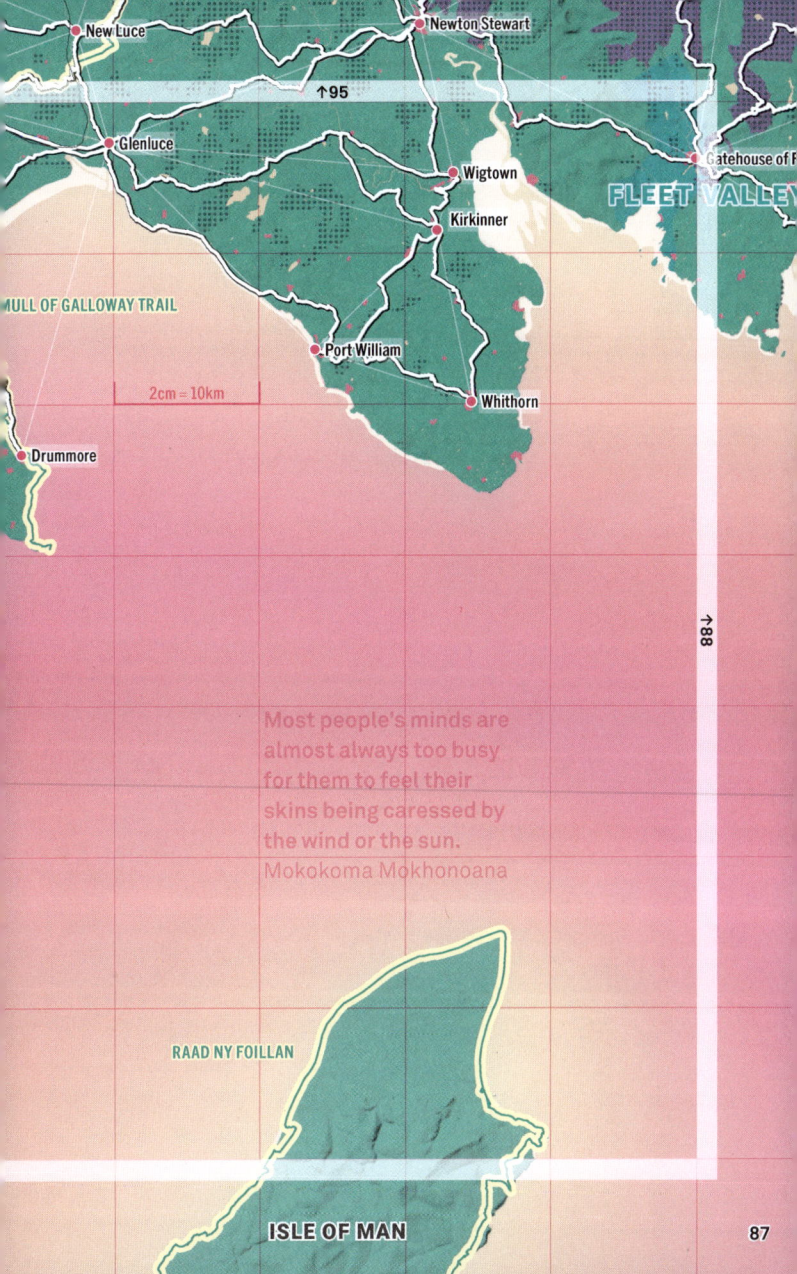

New Luce

Newton Stewart

↑95

Glenluce

Wigtown

Gatehouse of F

FLEET VALLEY

Kirkinner

MULL OF GALLOWAY TRAIL

2cm = 10km

Port William

Whithorn

Drummore

↑88

Most people's minds are
almost always too busy
for them to feel their
skins being caressed by
the wind or the sun.
Mokokoma Mokhonoana

RAAD NY FOILLAN

ISLE OF MAN

NITH ESTUARY

Castle Douglas

Dalbeattie
↑96

Gatehouse of Fleet

FLEET VALLEY

EAST STEWARTRY COAST

Kirkcudbright

2cm = 10km

Marypo[...]

Workington

↑87

What is
worth doing
at all is worth
doing slowly
and well.
Emma Bullet

Whitehaven

Cleator Moo[...]

Egremont

KING CHARLES III
ENGLAND COAST PATH

Ravenglass

↓80

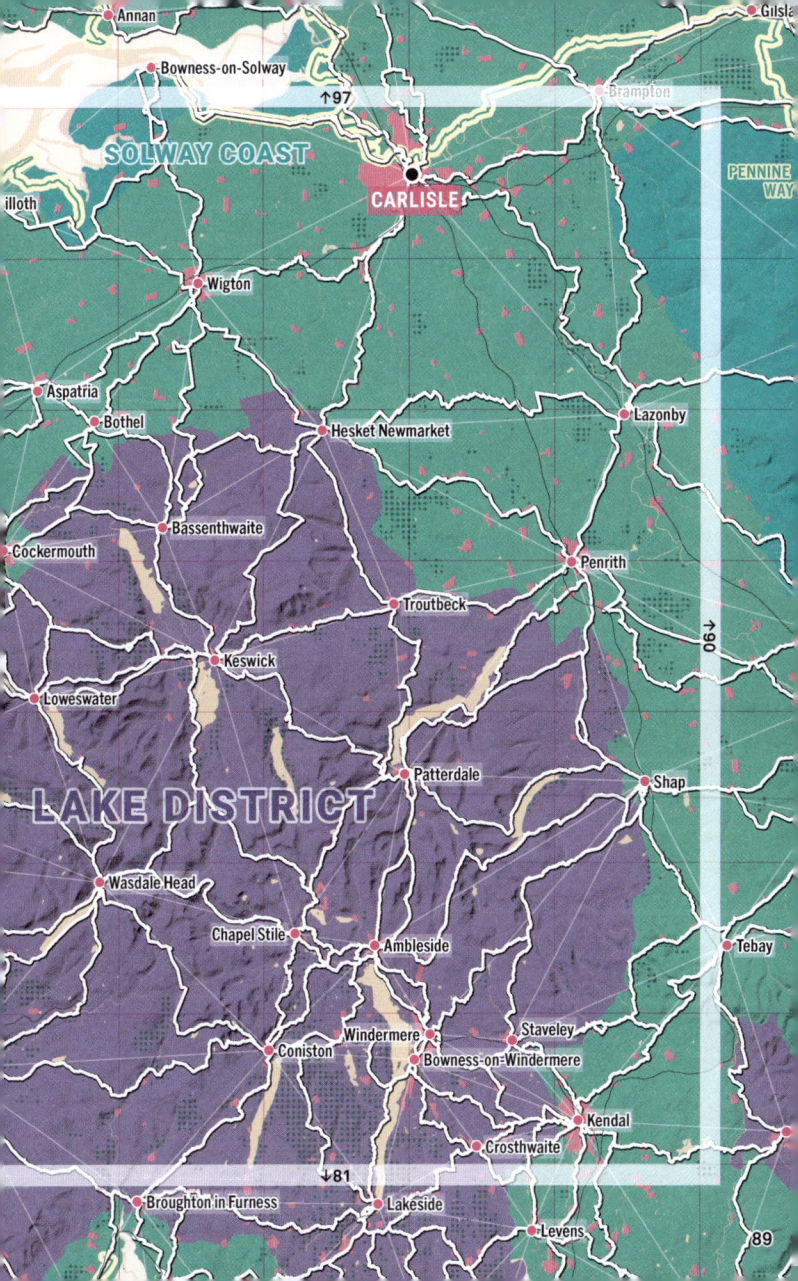

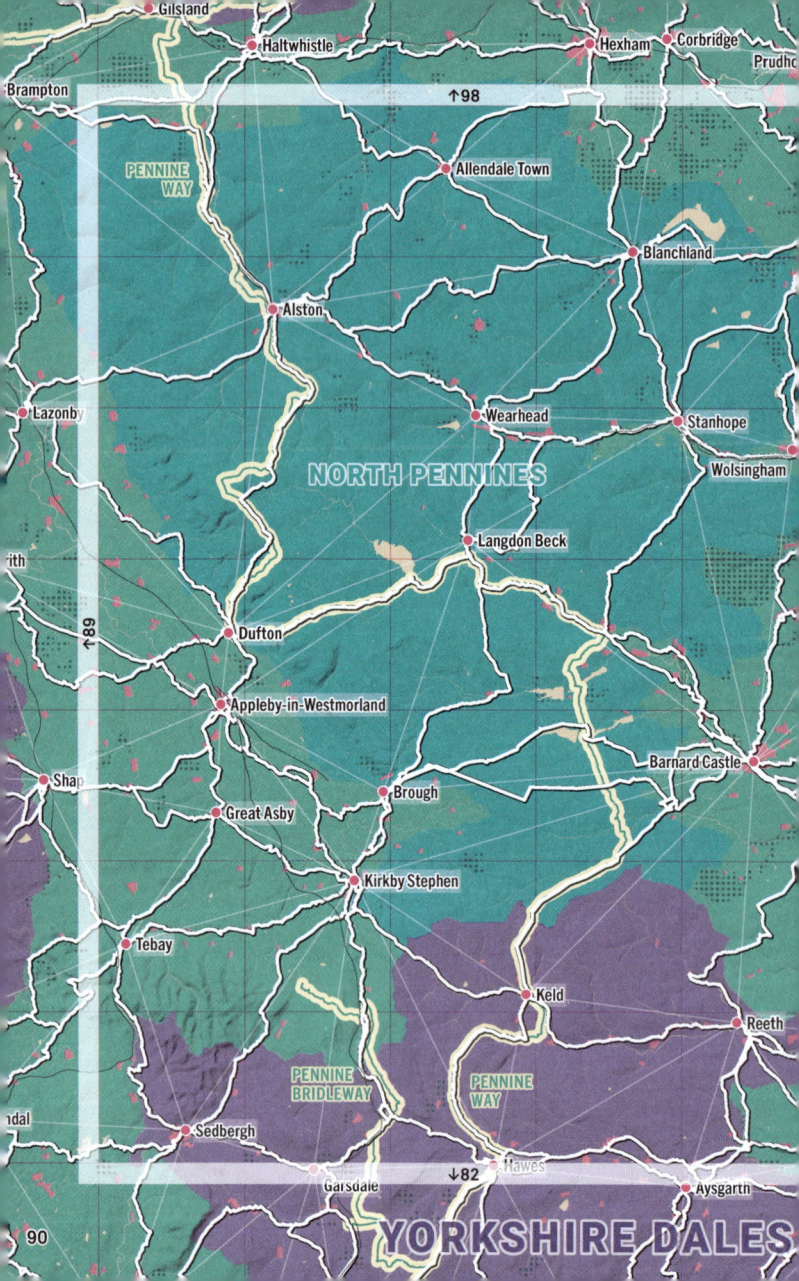

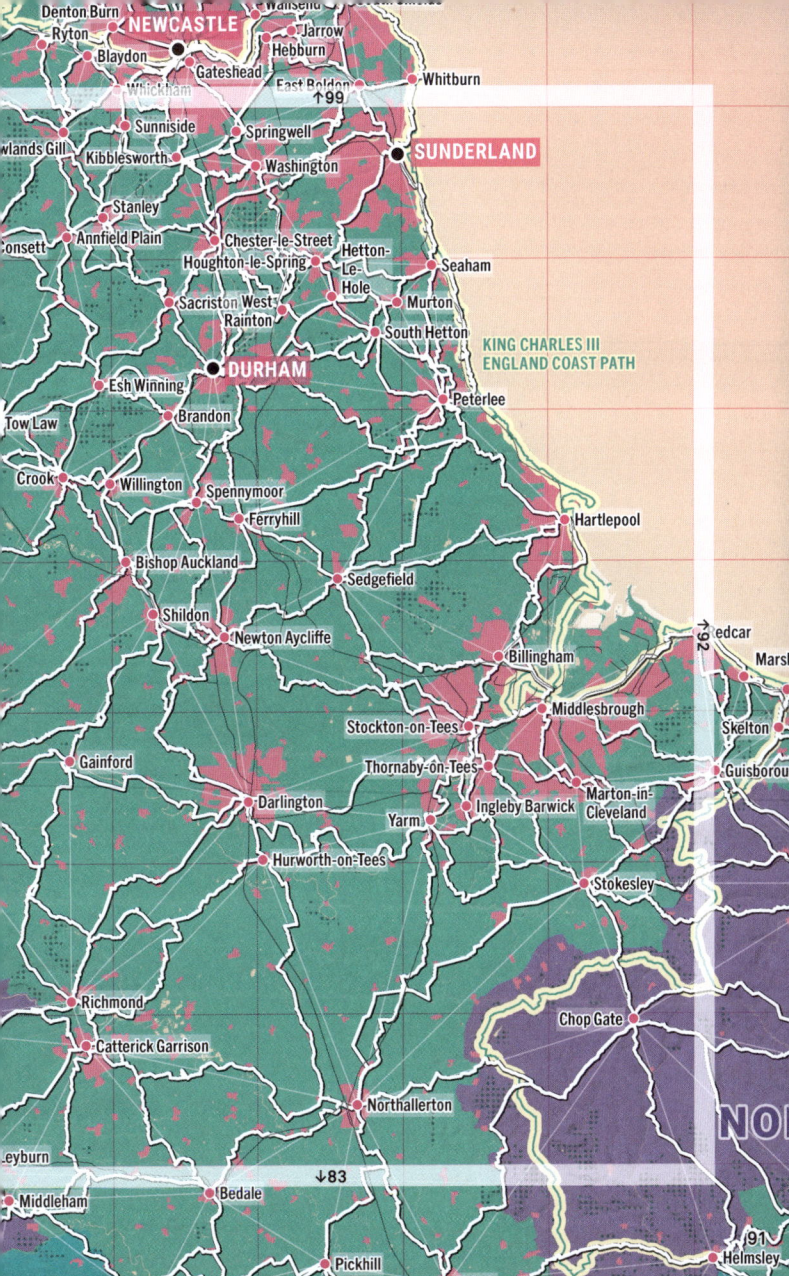

Denton Burn
Ryton
NEWCASTLE
Blaydon
Gateshead
Whickham
↑99
Wallsend
Hebburn
Jarrow
East Boldon
Whitburn

Sunniside
Springwell
rowlands Gill
Kibblesworth
Washington
SUNDERLAND

Stanley
onsett
Annfield Plain
Chester-le-Street
Houghton-le-Spring
Hetton-Le-Hole
Seaham
Sacriston
West Rainton
Murton
South Hetton

DURHAM
Esh Winning
Brandon
Peterlee

Tow Law
KING CHARLES III
ENGLAND COAST PATH

Crook
Willington
Spennymoor
Ferryhill
Hartlepool

Bishop Auckland
Sedgefield
↑92
edcar
Marsh

Shildon
Newton Aycliffe
Billingham
Skelton

Gainford
Stockton-on-Tees
Middlesbrough
Guisborou

Darlington
Thornaby-on-Tees
Ingleby Barwick
Marton-in-Cleveland
Yarm

Hurworth-on-Tees
Stokesley

Richmond
Chop Gate

Catterick Garrison

Northallerton
NO

eyburn
↓83

Middleham
Bedale
NOI

Pickhill
91
Helmsley,

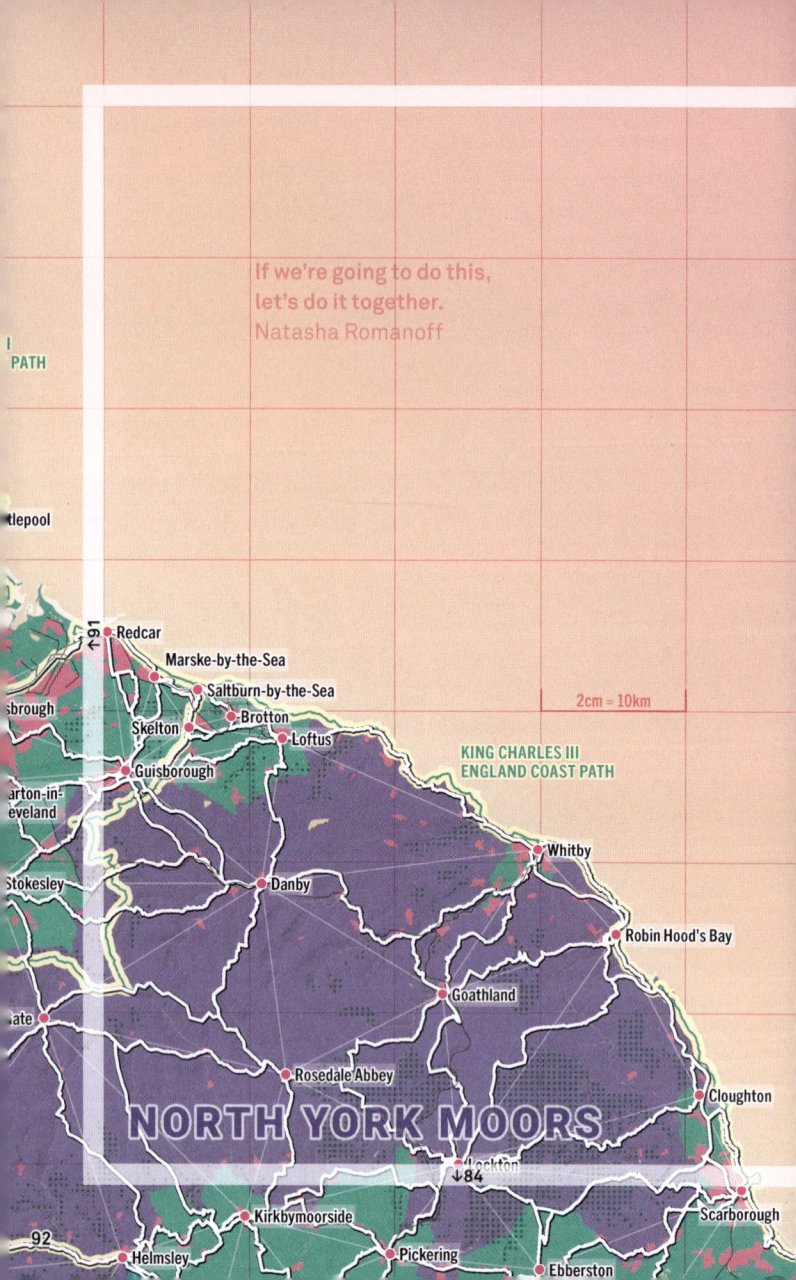

If we're going to do this,
let's do it together.
Natasha Romanoff

tlepool

↑91 Redcar
Marske-by-the-Sea
Saltburn-by-the-Sea
sbrough
Skelton   Brotton
Guisborough   Loftus

KING CHARLES III
ENGLAND COAST PATH

2cm = 10km

arton-in-
eveland

Stokesley

Danby   Whitby

Robin Hood's Bay

ate

Goathland

Rosedale Abbey

Cloughton

NORTH YORK MOORS

↓84 ockton

Kirkbymoorside   Scarborough

Helmsley   Pickering   Ebberston

You pray for rain, you
gotta deal with the mud
too. That's a part of it.
Coach Herman Boone

↓85

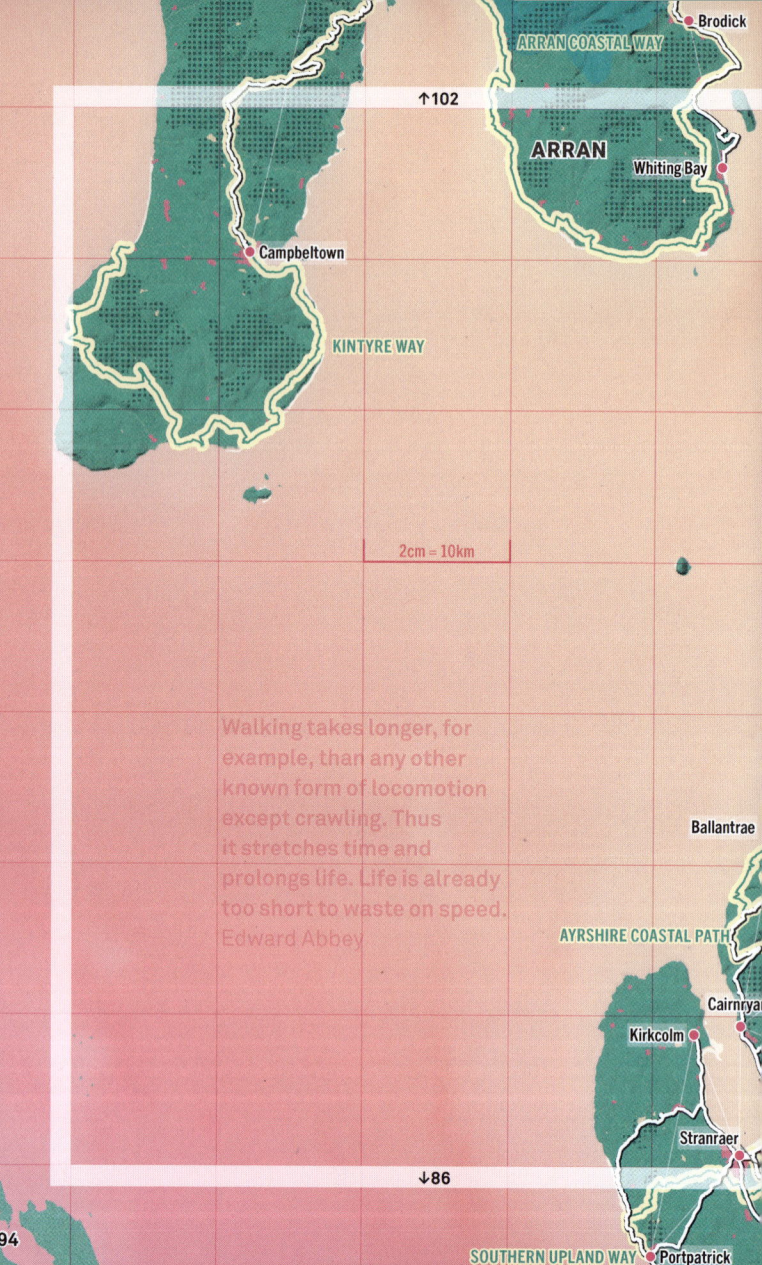

Brodick

ARRAN COASTAL WAY

ARRAN

Whiting Bay

Campbeltown

KINTYRE WAY

2cm = 10km

Walking takes longer, for
example, than any other
known form of locomotion
except crawling. Thus
it stretches time and
prolongs life. Life is already
too short to waste on speed.
Edward Abbey

Ballantrae

AYRSHIRE COASTAL PATH

Cairnryan

Kirkcolm

Stranraer

SOUTHERN UPLAND WAY   Portpatrick

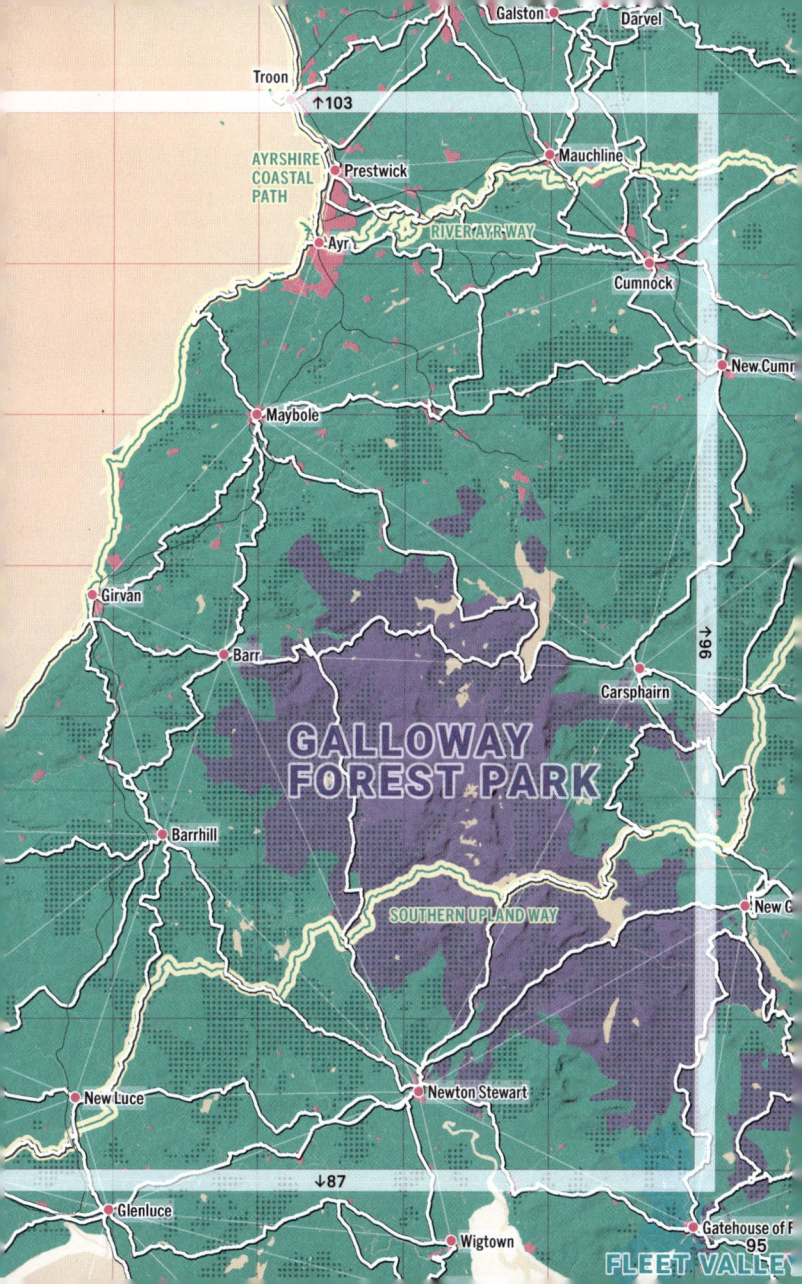

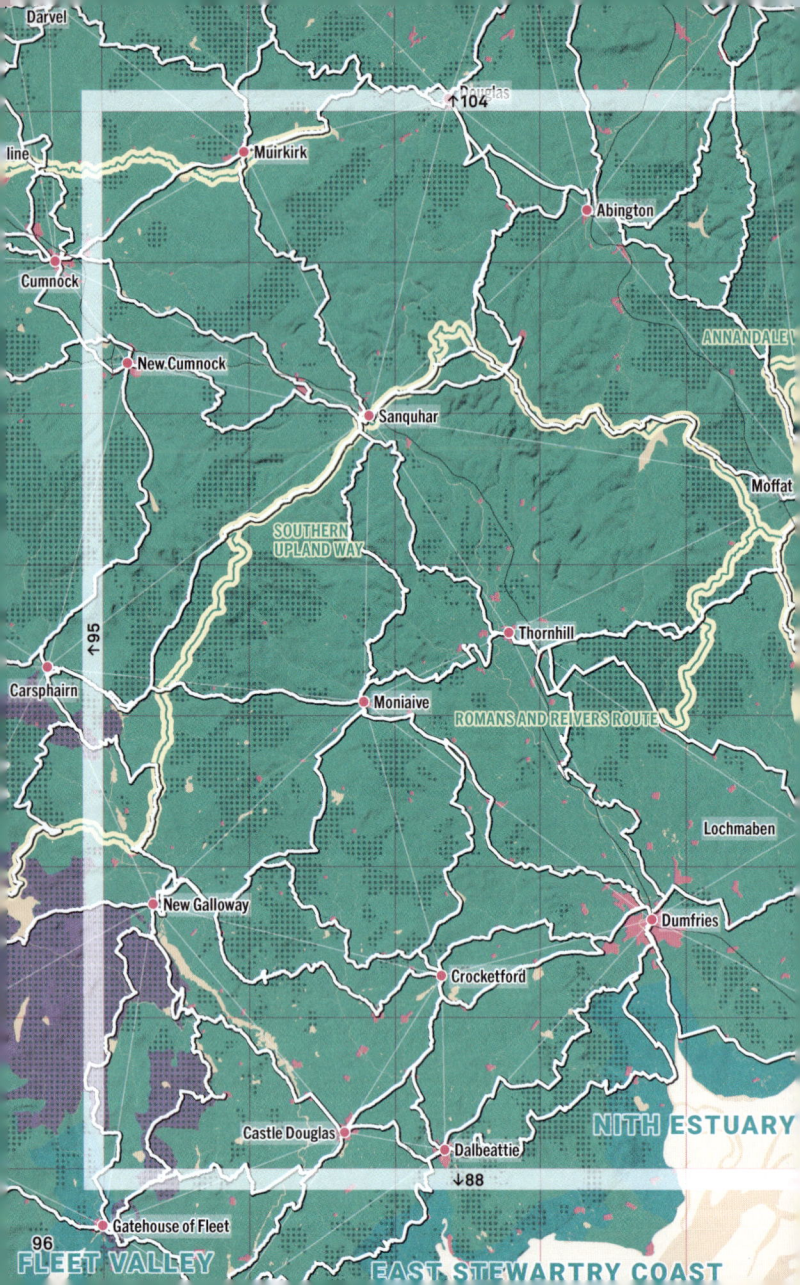

line

Cumnock

Muirkirk

↑104 Douglas

Abington

ANNANDALE V

New Cumnock

Sanquhar

Moffat

SOUTHERN
UPLAND WAY

↑95

Thornhill

Carsphairn

Moniaive

ROMANS AND REIVERS ROUTE

Lochmaben

New Galloway

Dumfries

Crocketford

Castle Douglas

Dalbeattie

NITH ESTUARY

↓88

Gatehouse of Fleet

FLEET VALLEY

EAST STEWARTRY COAST

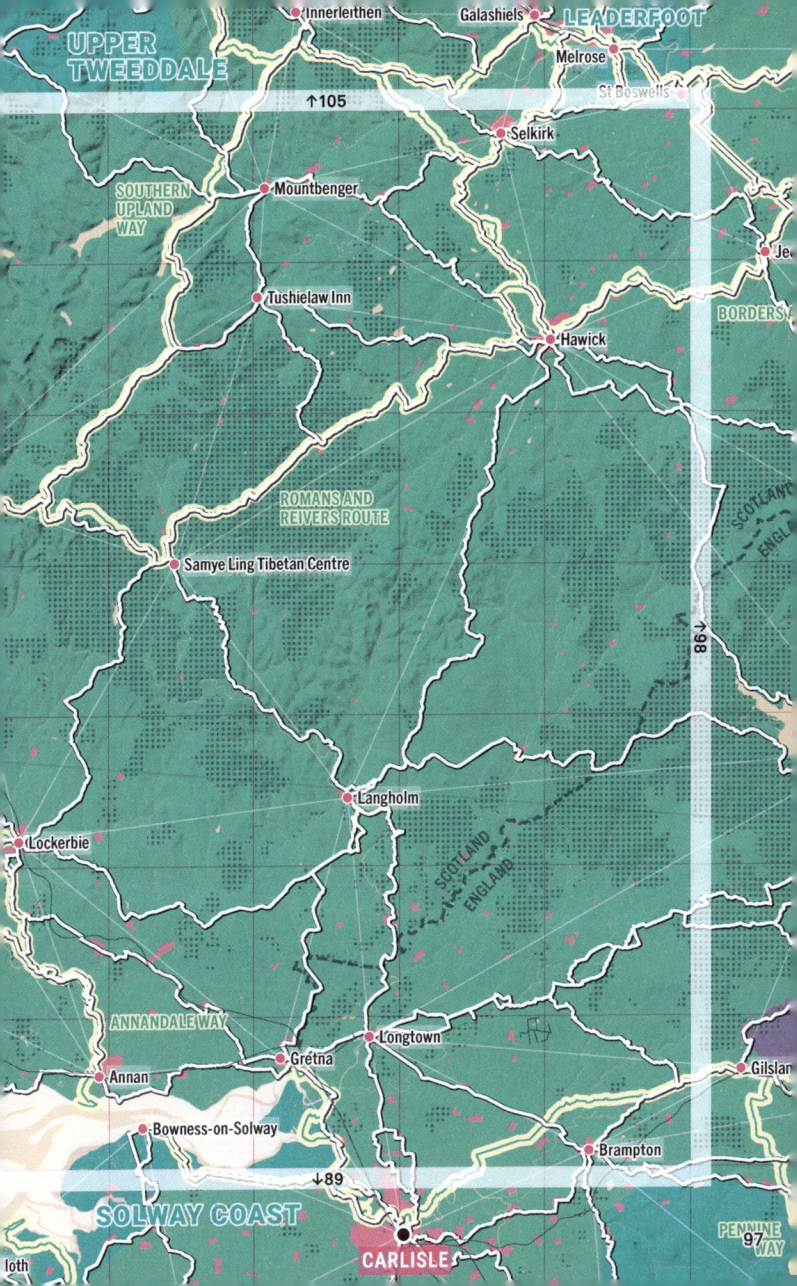

Innerleithen

Galashiels

LEADERFOOT

Melrose

St Boswells

↑105

Selkirk

SOUTHERN
UPLAND
WAY

Mountbenger

Je

BORDERS

Tushielaw Inn

Hawick

SCOTLAND
ENGL

ROMANS AND
REIVERS ROUTE

Samye Ling Tibetan Centre

↑98

Langholm

SCOTLAND
ENGLAND

Lockerbie

ANNANDALE WAY

L'ongtown

Gretna

Annan

Gilslan

Bowness-on-Solway

Brampton

↓89

SOLWAY COAST

PENNINE
WAY

loth

CARLISLE

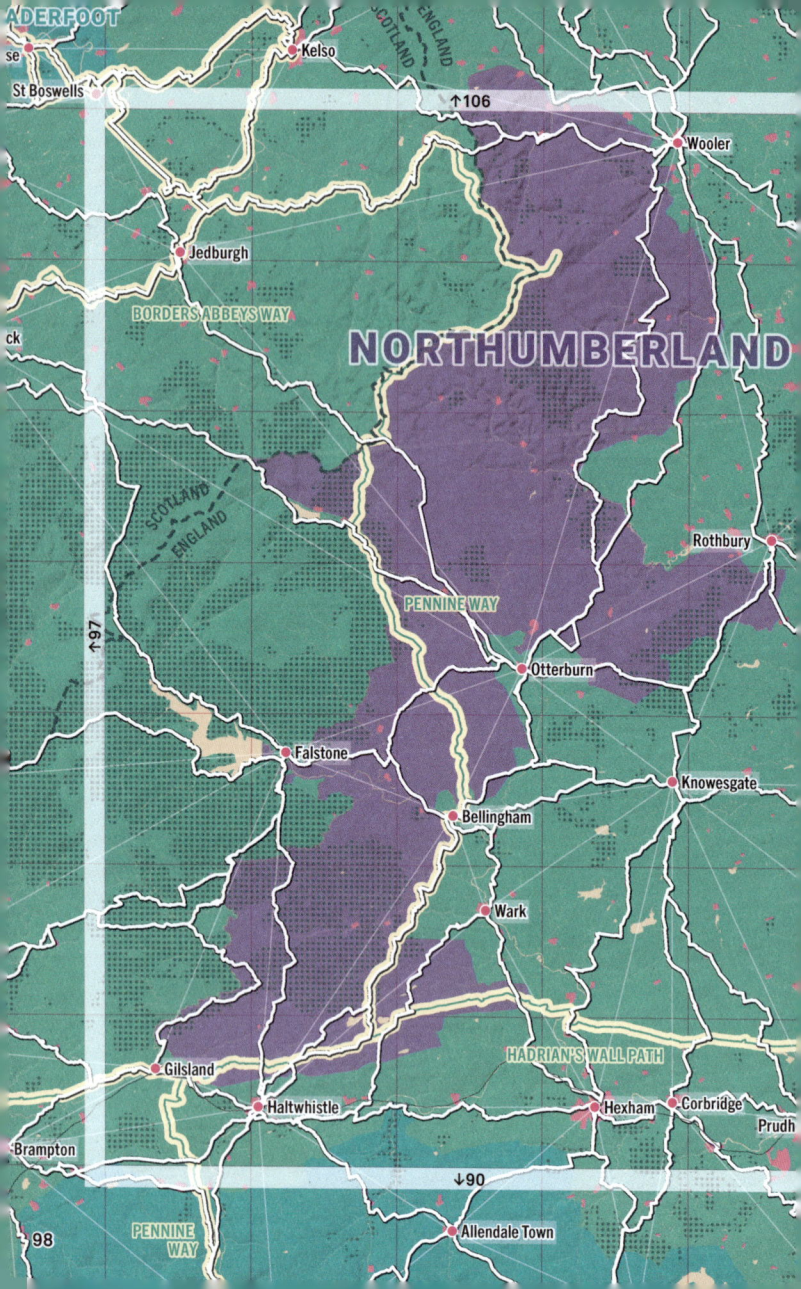

NORTHUMBERLAND COAST

↑107

KING CHARLES III
ENGLAND COAST PATH

Are you ready, boots?
Start walkin'
Nancy Sinatra

Alnwick

Amble

Ashington • Newbiggin-By-The-Sea
Morpeth • Stakeford
Bedlington • Blyth

2cm = 10km

Cramlington
Seaton Delaval
Ponteland • Wideopen • Whitley Bay
Shiremoor
Brunton Bridge • Tynemouth
Gosforth • Longbenton
Denton Burn • Wallsend • South Shields
Ryton • NEWCASTLE • Jarrow
Blaydon • Hebburn
Whickham • Gateshead • East Boldon • Whitburn
wlands Gill • Sunniside • Springwell
Kibblesworth • Washington • SUNDERLAND
Stanley

↓91

In an age of speed, I began
to think, nothing could
be more invigorating
than going slow. In an age
of distraction, nothing
can feel more luxurious
than paying attention.
Pico Iyer

SCAR
LUNG
GARV

COLONSAY

● Scalasaig

Ardluss

JURA

Port Askaig ●● Feolin Ferry

ISLAY

● Bridgend

↑102

● Portnahaven

● Port Ellen

2cm = 10km

↑110

Kilmartin

LOCH LOMOND AND
COWAL WAY

Ardlussa

Lochgilphead

KNAPDALE

Ormidale

KYLES OF BUTE

Rhubodach Ferry   Colintrai

Tighnabruaich

Portavadie

Tarbert

Rothesay

Kennacraig

KINTYRE WAY

↑101

Claonaig

WEST ISLAND WAY

Lochranza

NORTH ARRAN

Brodick

ARRAN COASTAL WAY

↓94

ARRAN

Whiting Bay

Tarbet

↑111

Aberfoyle

Rowardennan

LOCH
LOMOND

Garelochhead

THREE LOCHS WAY

Drymen

Helensburgh

WEST HIGHLAND WAY

Kilcreggan

Alexandria

Strathblane

Milton of Ca

Gourock

JOHN MUIR WAY

Dunoon

Greenock

Dumbarton

Milngavie

Port Glasgow

Old Kilpatrick

Bearsden

Bishopton

Clydebank

Bishopbrig

Kilmacolm

Erskine

Kelvindale

Stepps

Wemyss Bay

Renfrew

Gilshochill

Strepps

Bridge of Weir

GLASGOW
GLASCHU

Paisley

Pollokshields

Shettlesto

Johnstone

Rutherglen

Ud

Cumbrae Ferry

Thornliebank

Largs

Lochwinnoch

Barrhead

Cambuslang

Giffnock

Millport

Uplawmoor

Busby

Kilbirnie

Beith

East

West Kilbride

Dalry

Whitelee
Wind Farm

AYRSHIRE
COASTAL PATH

Stewarton

Ardrossan

Kilwinning

Stevenston

Fenwick

2cm = 10km

Saltcoats

Irvine

Kilmarnock Newmilns

Galston

Darvel

Troon

↓95

Mauchline

AYRSHIRE
COASTAL
PATH

Prestwick

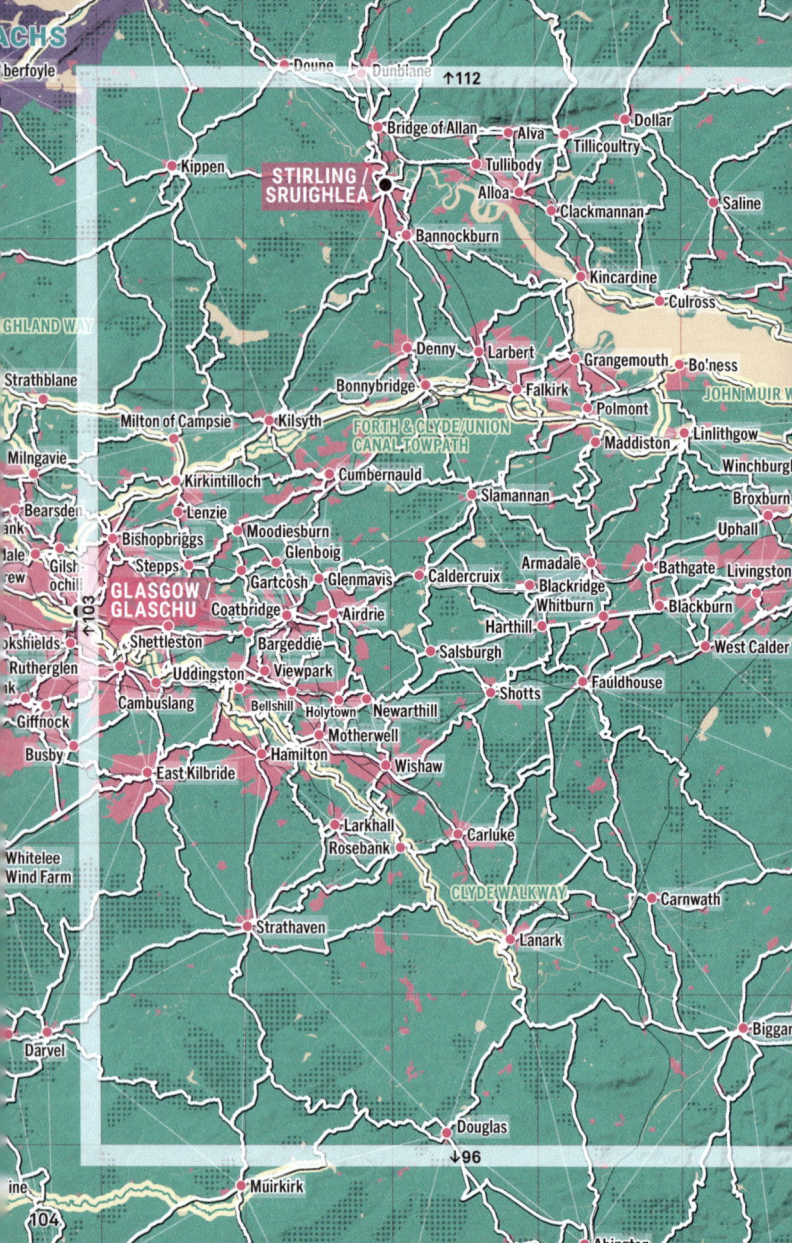

ACHS

berfoyle

Doune   Dunblane

Kippen   Bridge of Allan   Alva   Dollar

STIRLING /   Tillicoultry
SRUIGHLEA   Tullibody

Alloa   Saline
Clackmannan

Bannockburn

Kincardine

Culross

GHLAND WAY

Strathblane   Denny   Larbert   Grangemouth   Bo'ness

Bonnybridge   Falkirk   JOHN MUIR W

Milton of Campsie   Kilsyth   Polmont

Milngavie   FORTH & CLYDE/UNION   Linlithgow
CANAL TOWPATH   Maddiston

Kirkintilloch   Cumbernauld   Winchburgh

Bearsden   Lenzie   Broxburn

nk   Slamannan   Uphall

dale   Moodiesburn
Gilsh   Bishopbriggs   Glenboig   Armadale   Bathgate   Livingston
ew   ochil   Stepps   Gartcosh   Glenmavis   Caldercruix   Blackridge   Blackburn
GLASGOW /   Coatbridge   Airdrie   Whitburn   West Calder
GLASCHU   Harthill
kshields   Shettleston   Bargeddie   Salsburgh   Fauldhouse
Rutherglen   Uddingston   Viewpark   Shotts
k   Cambuslang   Bellshill   Holytown   Newarthill
Giffnock   Motherwell
Busby   Hamilton   Wishaw

East Kilbride

Larkhall   Carluke
Rosebank

Whitelee   CLYDE WALKWAY   Carnwath
Wind Farm

Strathaven   Lanark

Biggar

Darvel

Douglas

Muirkirk

ine

Abington

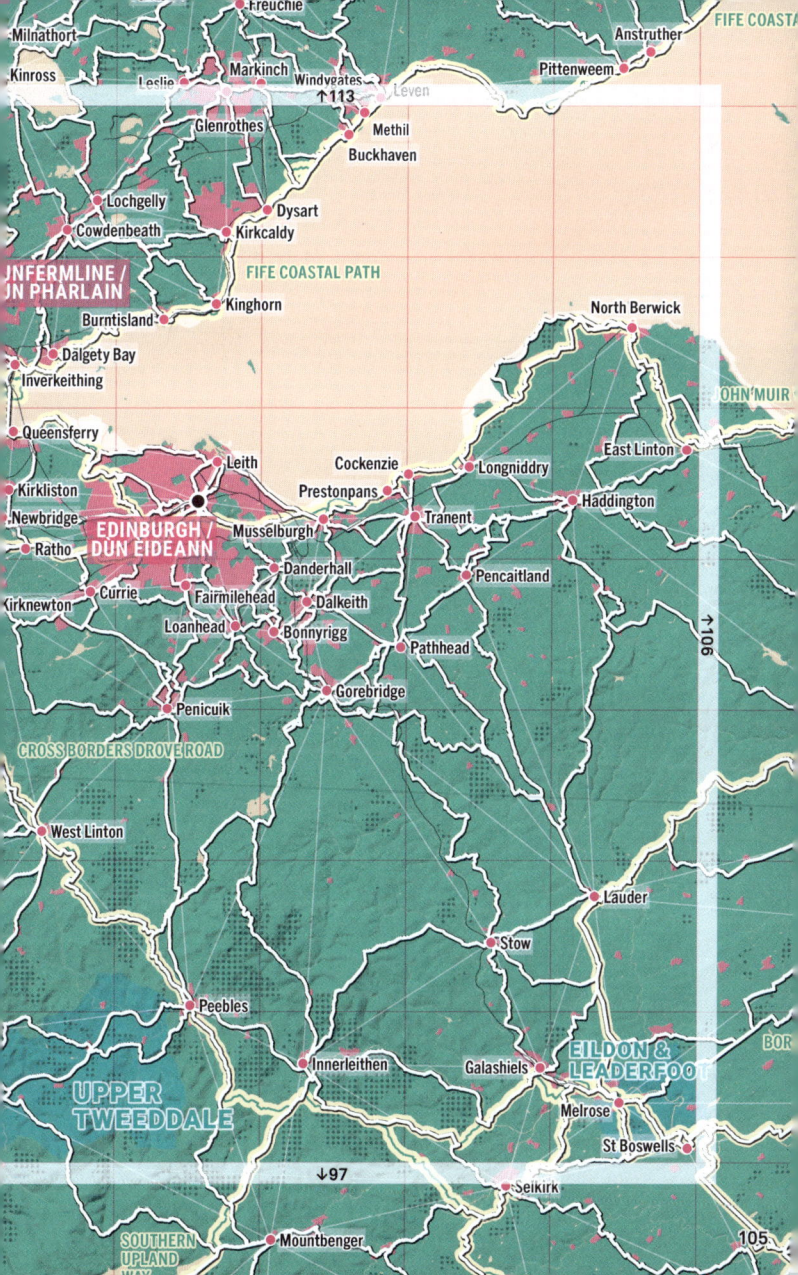

Milnathort
Kinross
Leslie
Markinch
Windygates
↑113
Freuchie
Anstruther
Pittenweem
Leven
FIFE COASTAL
Glenrothes
Methil
Buckhaven
Lochgelly
Dysart
Cowdenbeath
Kirkcaldy
NFERMLINE /
N PHARLAIN
FIFE COASTAL PATH
Burntisland
Kinghorn
North Berwick
Dalgety Bay
Inverkeithing
OHN'MUIR
Queensferry
Kirkliston
East Linton
Newbridge
Cockenzie
Longniddry
Haddington
Ratho
Leith
Prestonpans
EDINBURGH /
DUN EIDEANN
Musselburgh
Tranent
Kirknewton
Currie
Danderhall
Pencaitland
Fairmilehead
Dalkeith
Loanhead
Bonnyrigg
Pathhead
Gorebridge
↑106
Penicuik
CROSS BORDERS DROVE ROAD
West Linton
Lauder
Stow
UPPER
TWEEDDALE
Peebles
Innerleithen
Galashiels
EILDON &
LEADERFOOT
BOR
Melrose
St Boswells
↓97
Selkirk
SOUTHERN
UPLAND
Mountbenger
105

FIFE COASTAL PATH

Anstruther

↑114

In every job that must be done,
there is an element of fun.
Mary Poppins

North Berwick

JOHN MUIR WAY

Dunbar

East Linton

ddington

↑105

Grantshouse

Eyemouth

BERWICKSHIRE
COASTAL PATH

SOUTHERN
UPLAND WAY

Duns

Berwick-
upon-
Tweed

Lauder

Greenlaw

Norham

DON &
ADERFOOT

BORDERS ABBEYS WAY

Coldstream

Etal

ENGLAND
SCOTLAND

se

Kelso

St Boswells

↓98

Wooler

2cm = 10km

Now, bring me that horizon.
Captain Jack Sparrow

KING CHARLES III
ENGLAND COAST PATH

NORTHUMBERLAND
COAST

TIREE

Scarinish

I walk for miles along the highway
Well, that's just my way
Of sayin' I love you
Patsy Cline

Kilchoan

COLL

Arinagour

Tobermory

Salen

F

↑110 MULL

LOCH NA KEAL,
ISLE OF MULL

Fionnphort

Carsaig

2cm = 10km

SCAR
LUNGA
GARV

COLONSAY

Scalasaig

Camusnagaul Fe

↑118

Acharacle

Salen

Strontian

Ardgour Ferry   Corran Fe

Ballachulish

ermory

Lochaline

LYNN OF LORN

Salen

Fishnish Ferry

↑109

Craignure

MULL

Oban

Taynuilt

Carsaig

Kilninver

Melfort

SCARBA,
LUNGA & THE
GARVELLACHS

↓102
Kilmartin

LOCH LOMOND AN
GOWAL WAY

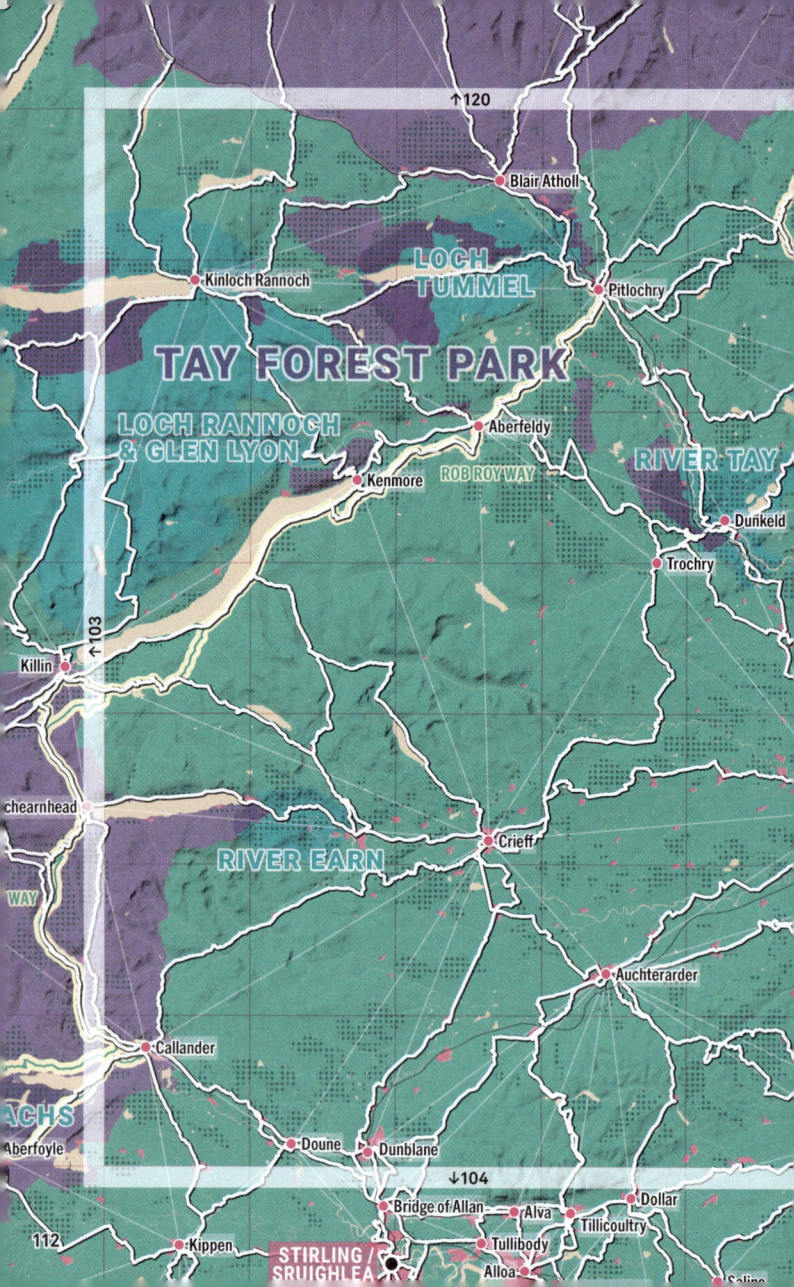

Blair Atholl

Kinloch Rannoch

LOCH TUMMEL

Pitlochry

TAY FOREST PARK

LOCH RANNOCH & GLEN LYON

Aberfeldy

Kenmore

ROB ROY WAY

RIVER TAY

Dunkeld

Trochry

↑103

Killin

chearnhead

RIVER EARN

Crieff

WAY

Auchterarder

Callander

Doune

Dunblane

Aberfoyle

ACHS

112

Kippen

Bridge of Allan

Alva

Dollar

Tillicoultry

Tullibody

STIRLING / SRUIGHLEA

Alloa

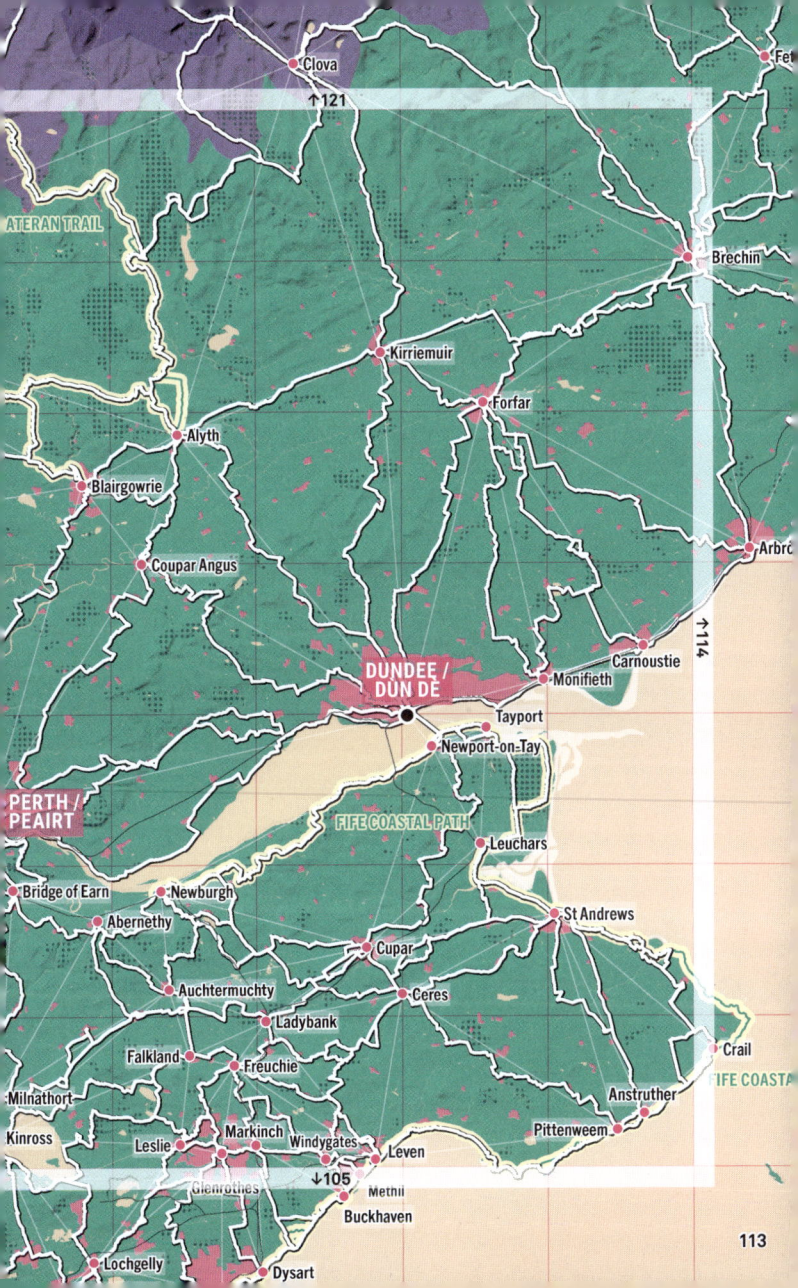

ATERAN TRAIL

Clova

Fet

Brechin

Kirriemuir

Forfar

Alyth

Blairgowrie

Coupar Angus

Arbr

↑114

Carnoustie

DUNDEE / DÙN DÈ

Monifieth

Tayport

Newport-on-Tay

PERTH / PEAIRT

FIFE COASTAL PATH

Leuchars

Bridge of Earn

Newburgh

Abernethy

St Andrews

Cupar

Auchtermuchty

Ceres

Crail

Ladybank

FIFE COASTA

Falkland

Freuchie

Milnathort

Anstruther

Kinross

Pittenweem

Leslie

Markinch

Windygates

Leven

Lochgelly

Glenrothes

Methil

Buckhaven

Dysart

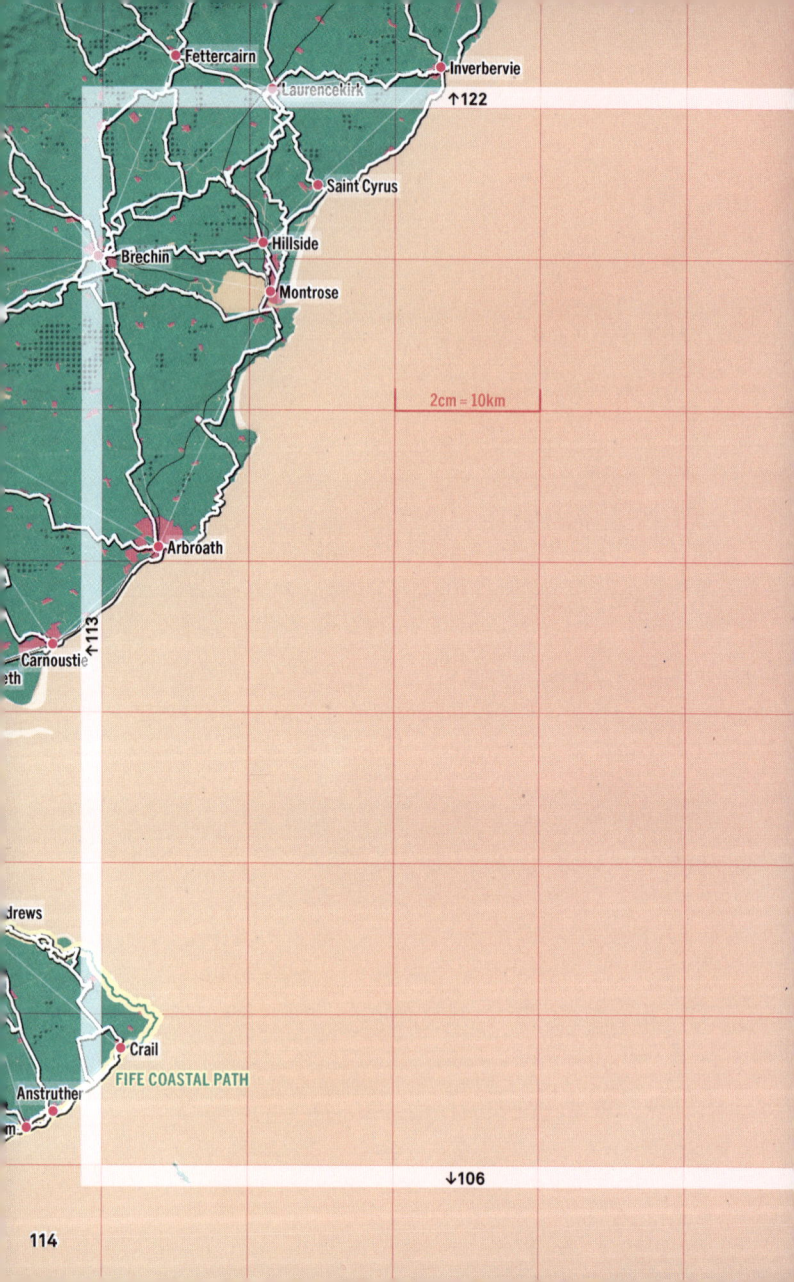

Fettercairn

Inverbervie

Laurencekirk

↑122

Saint Cyrus

Hillside

Brechin

Montrose

2cm = 10km

Arbroath

↑113

Carnoustie

eth

drews

Crail

FIFE COASTAL PATH

Anstruther

m

↓106

↑123

You're off the edge of the map, mate. Here there be monsters.
Captain Barbossa

↓107

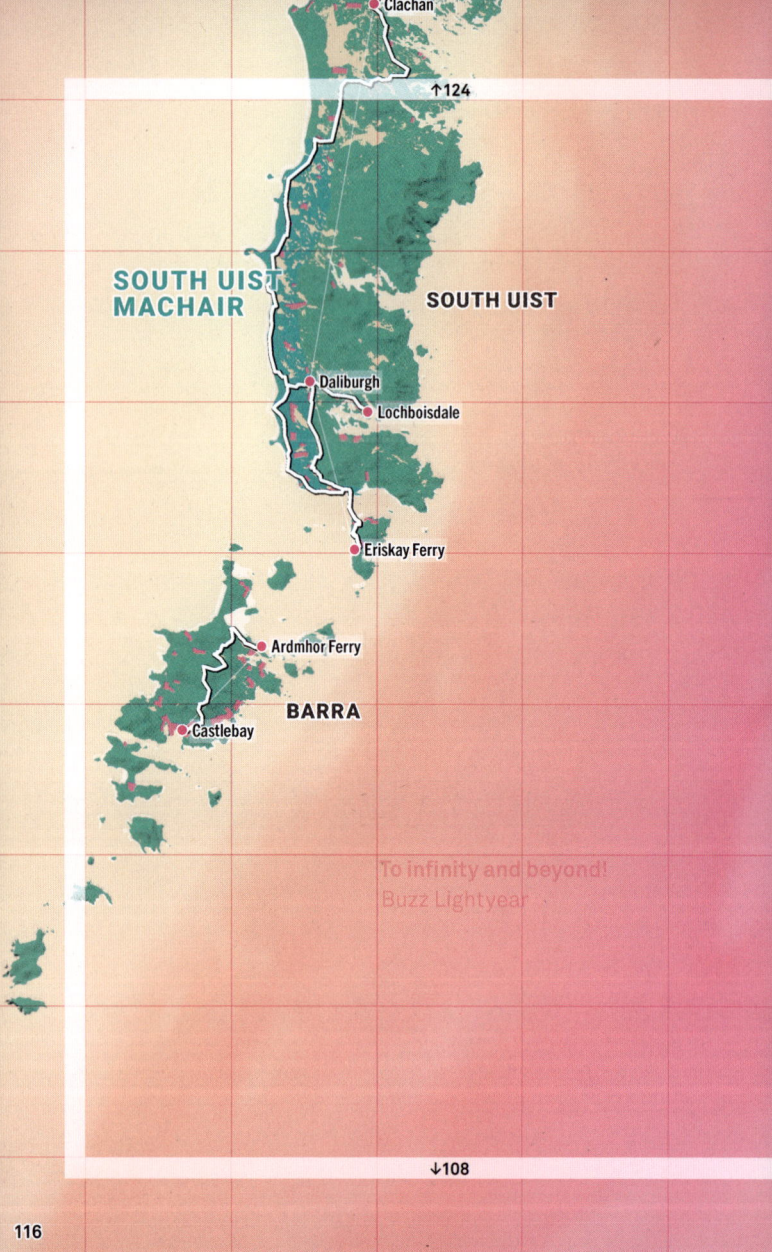

Clachan

SOUTH UIST
MACHAIR

SOUTH UIST

Daliburgh

Lochboisdale

Eriskay Ferry

Ardmhor Ferry

BARRA

Castlebay

To infinity and beyond!
Buzz Lightyear

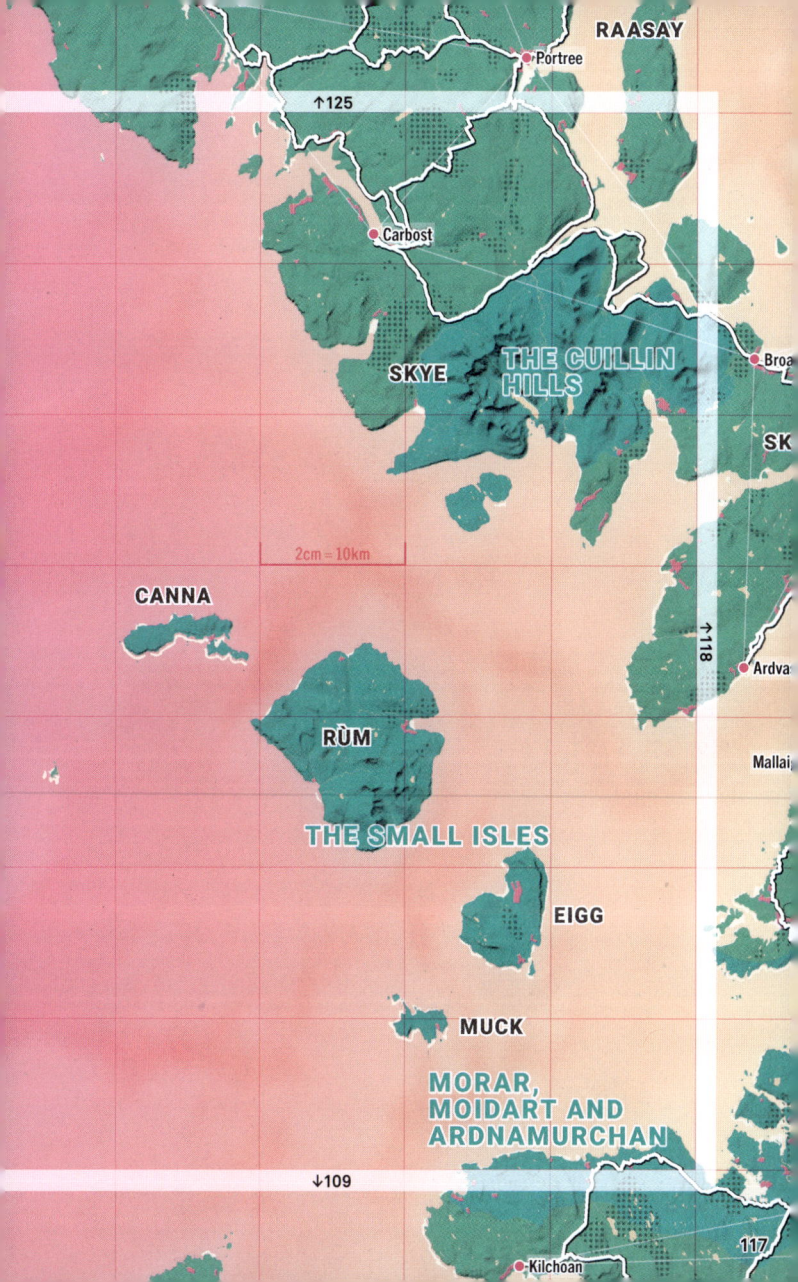

RAASAY

Portree

↑125

Carbost

THE CUILLIN
HILLS

Broa

SKYE

SK

2cm = 10km

CANNA

↑118

Ardva

RÙM

Mallai

THE SMALL ISLES

EIGG

MUCK

MORAR,
MOIDART AND
ARDNAMURCHAN

↓109

117

Kilchoan

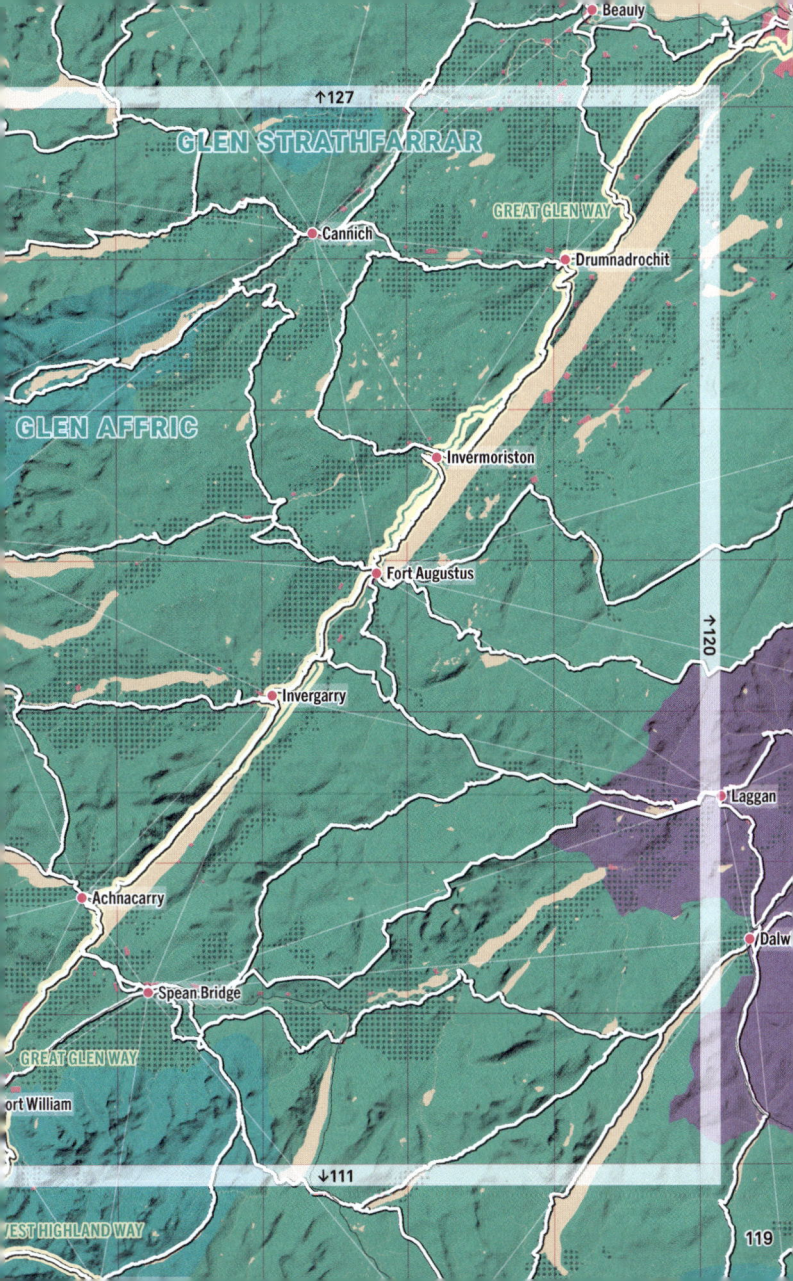

GLEN STRATHFARRAR

GREAT GLEN WAY

● Cannich

● Drumnadrochit

GLEN AFFRIC

● Invermoriston

● Fort Augustus

● Invergarry

● Laggan

● Achnacarry

● Dalw

● Spean Bridge

GREAT GLEN WAY

ort William

● Beauly

EST HIGHLAND WAY

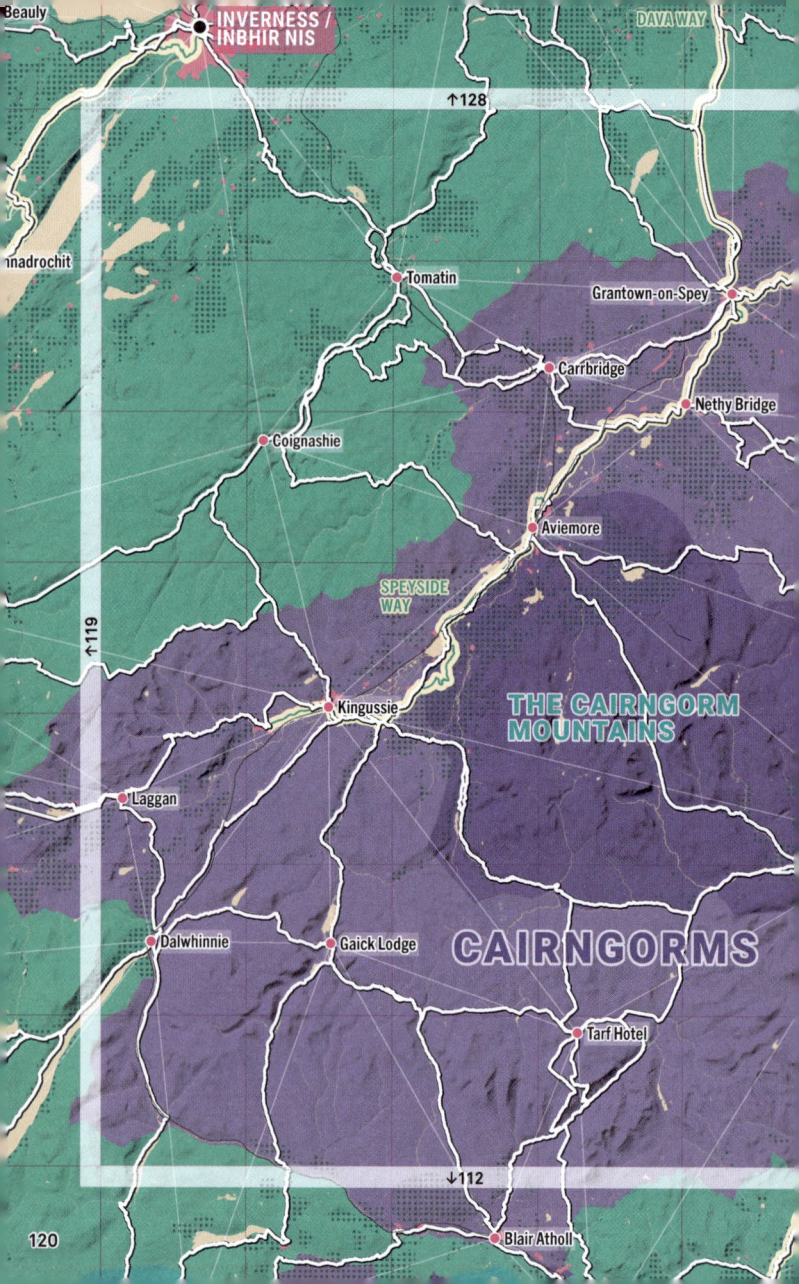

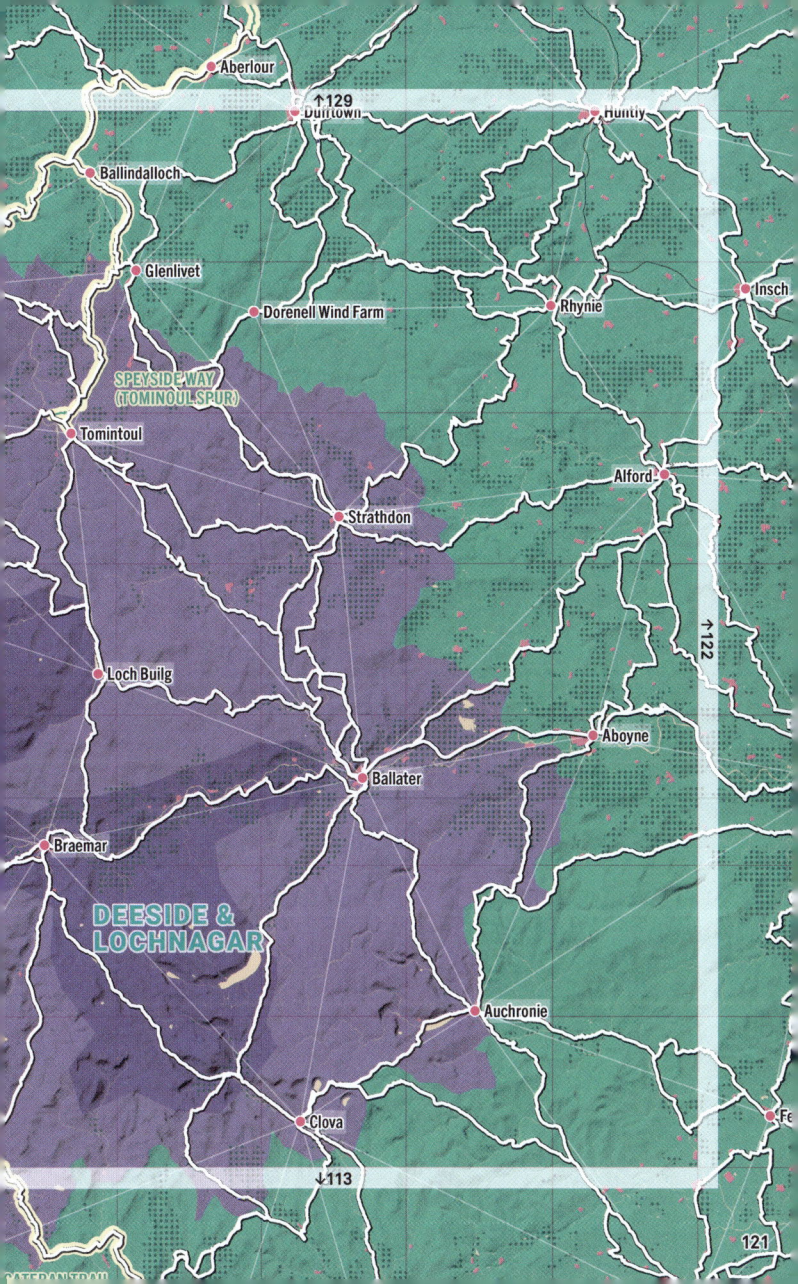

Aberlour

Dufftown

Huntly

Ballindalloch

Glenlivet

Insch

Dorenell Wind Farm

Rhynie

SPEYSIDE WAY
(TOMINOUL SPUR)

Tomintoul

Alford

Strathdon

↑122

Loch Builg

Aboyne

Ballater

Braemar

DEESIDE &
LOCHNAGAR

Auchronie

Fe

Clova

Huntly

Methlick

Rothienorman

Ellon

Insch

Oldmeldrum

Newburgh

Inverurie

Newmachar

Alford

Kintore

Balmedie

Monymusk

Blackburn

FORMARTINE AND
BUCHAN WAY

Westhill

ABERDEEN /
OBAR DHEATHAIN

Aboyne

Peterculter

Banchory

Portlethen

Stonehaven

2cm = 10km

Fettercairn

Inverbervie

Laurencekirk

Saint Cyrus

You have to act as if it were possible
to radically transform the world. And
you have to do it all the time.
Angela Davis

↓115

Life shrinks or expands in
proportion to one's courage.
Anaïs Nin

Leverburgh

Berneray Ferry

Malacleit

NORTH UIST

Lochmaddy

Carinish

Clachan

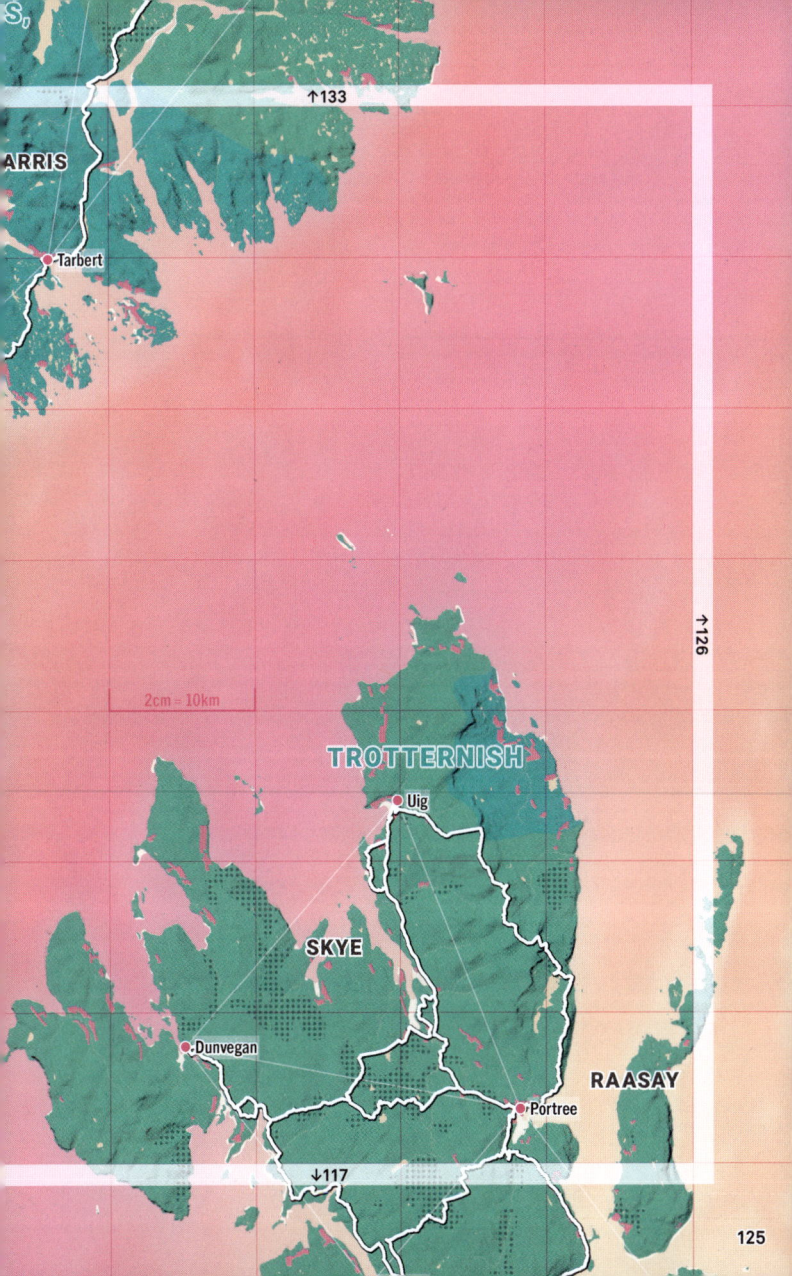

ARRIS

●·Tarbert

2cm = 10km

TROTTERNISH

●·Uig

SKYE

●·Dunvegan

RAASAY

●·Portree

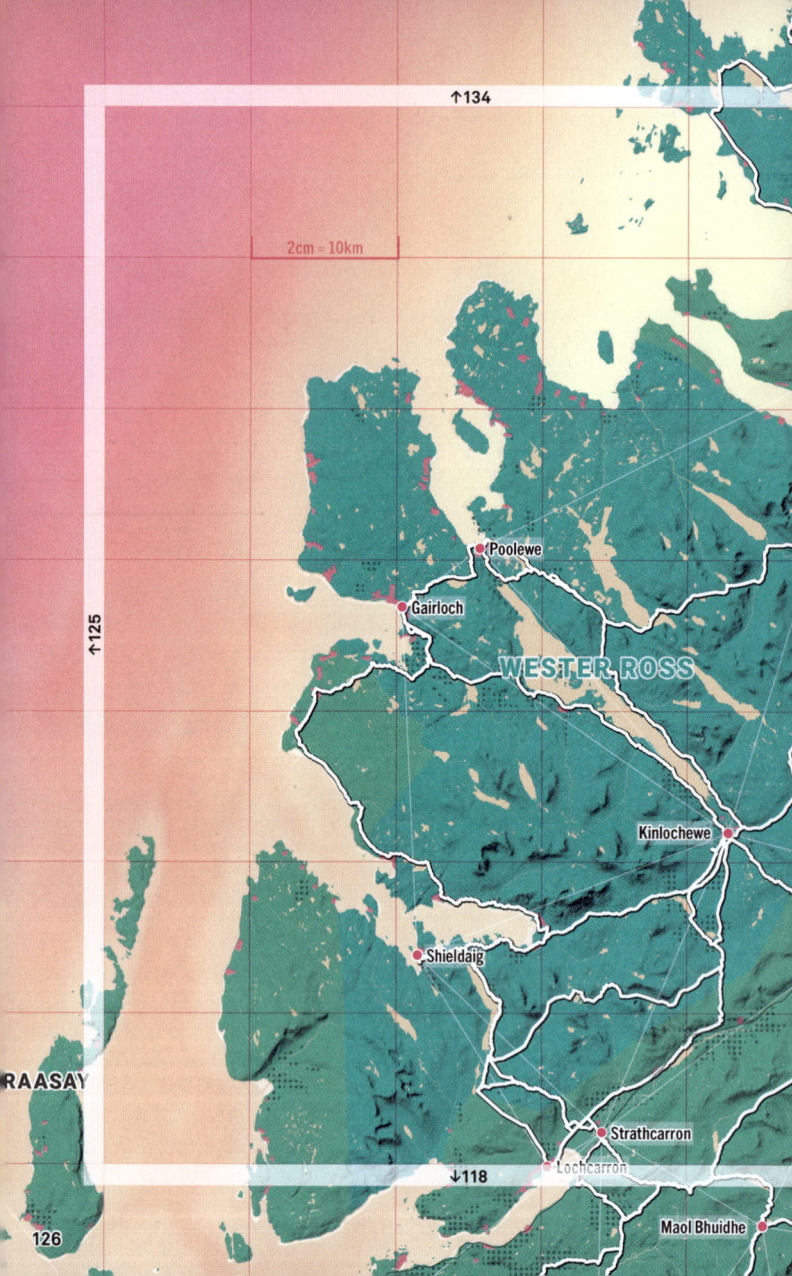

2cm = 10km

Poolewe

Gairloch

WESTER ROSS

Kinlochewe

Shieldaig

RAASAY

Strathcarron

Lochcarron

Maol Bhuidhe

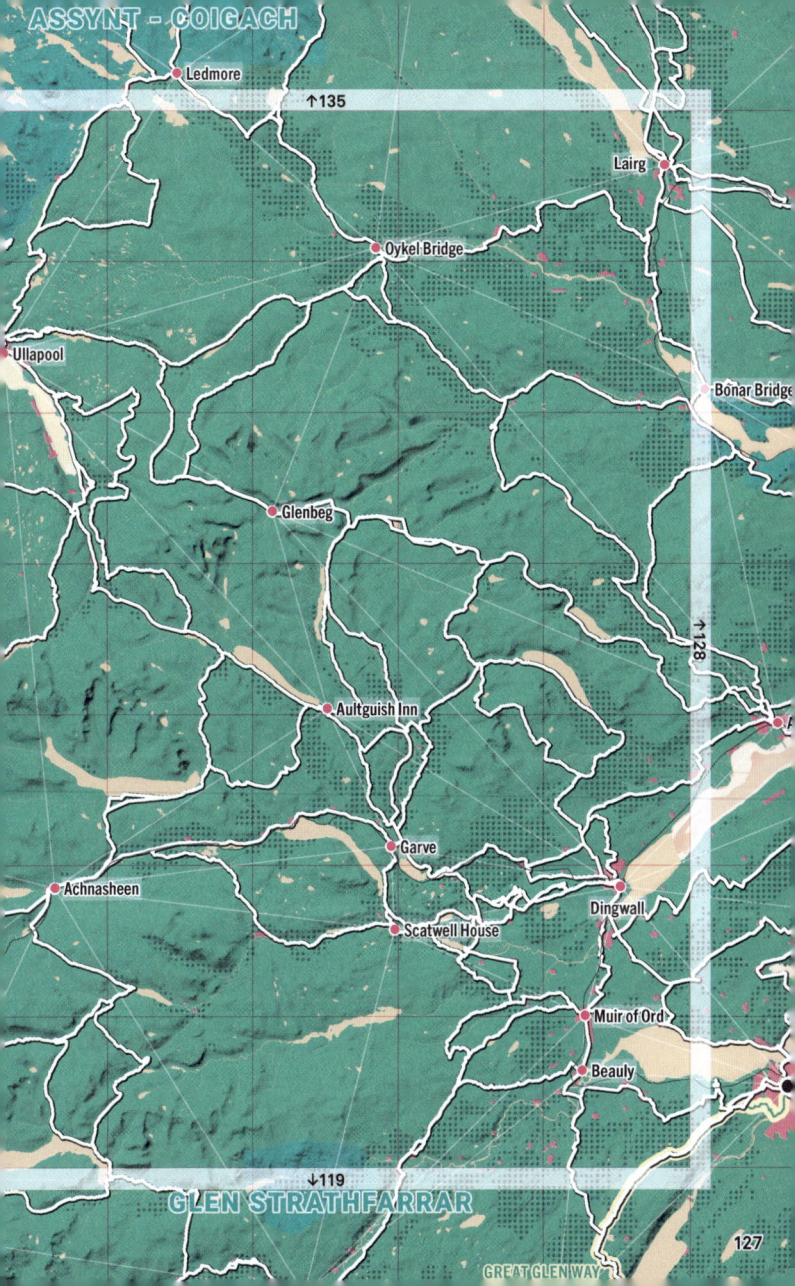

Ledmore

↑135

Lairg

Oykel Bridge

Ullapool

Bonar Bridge

Glenbeg

↑128

Aultguish Inn

Garve

Achnasheen

Dingwall

Scatwell House

Muir of Ord

Beauly

↓119

GREAT GLEN WAY

Helmsdale

Lairg

Golspie

Bonar Bridge

Dornoch

DORNOCH FIRTH

Tain

Alness

Invergordon

Nigg Ferry

Cromarty

Dingwall

Forres

Fortrose

Nairn

Muir of Ord

Beauly

Cawdor

**INVERNESS /
INBHIR NIS**

DAVA WAY

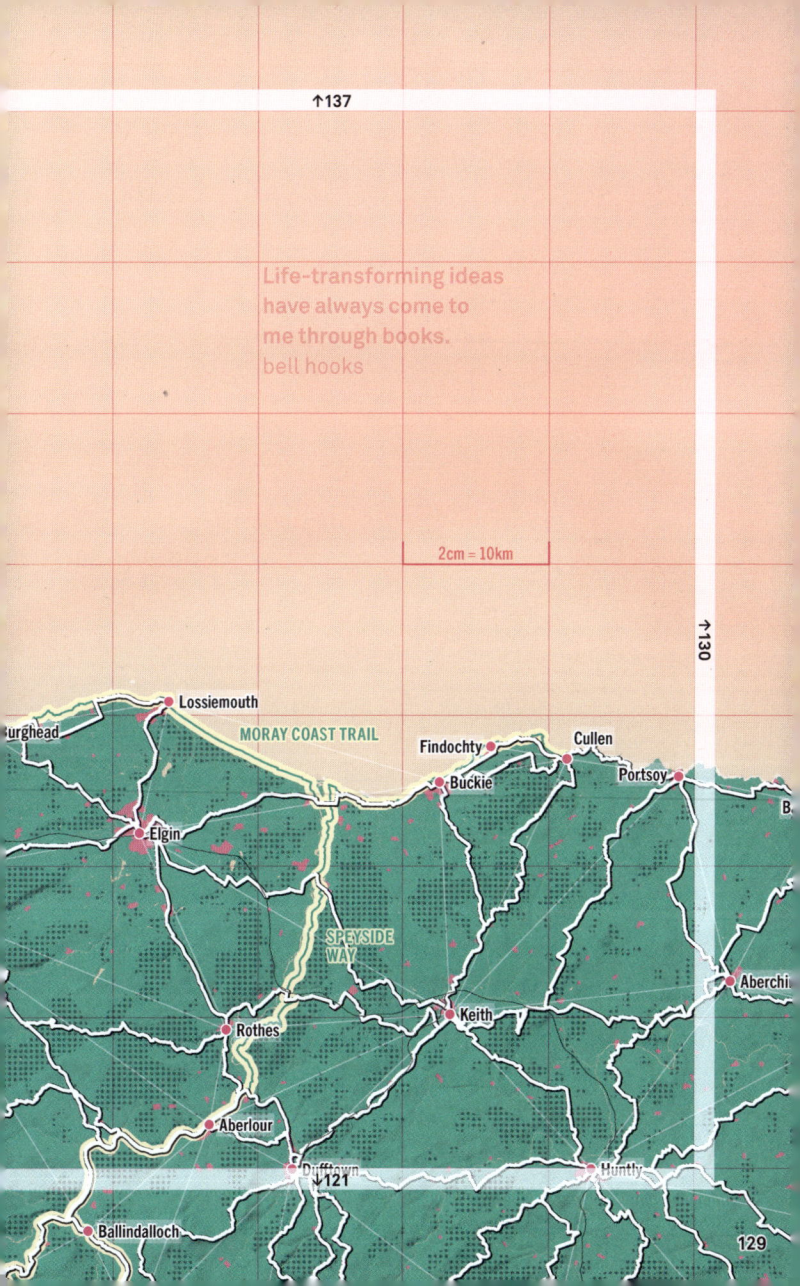

Life-transforming ideas
have always come to
me through books.
bell hooks

2cm = 10km

Lossiemouth

urghead

MORAY COAST TRAIL

Findochty

Cullen

Buckie

Portsoy

B.

Elgin

SPEYSIDE
WAY

Aberchi

Rothes

Keith

Aberlour

Ballindalloch

Dufftown

Huntly

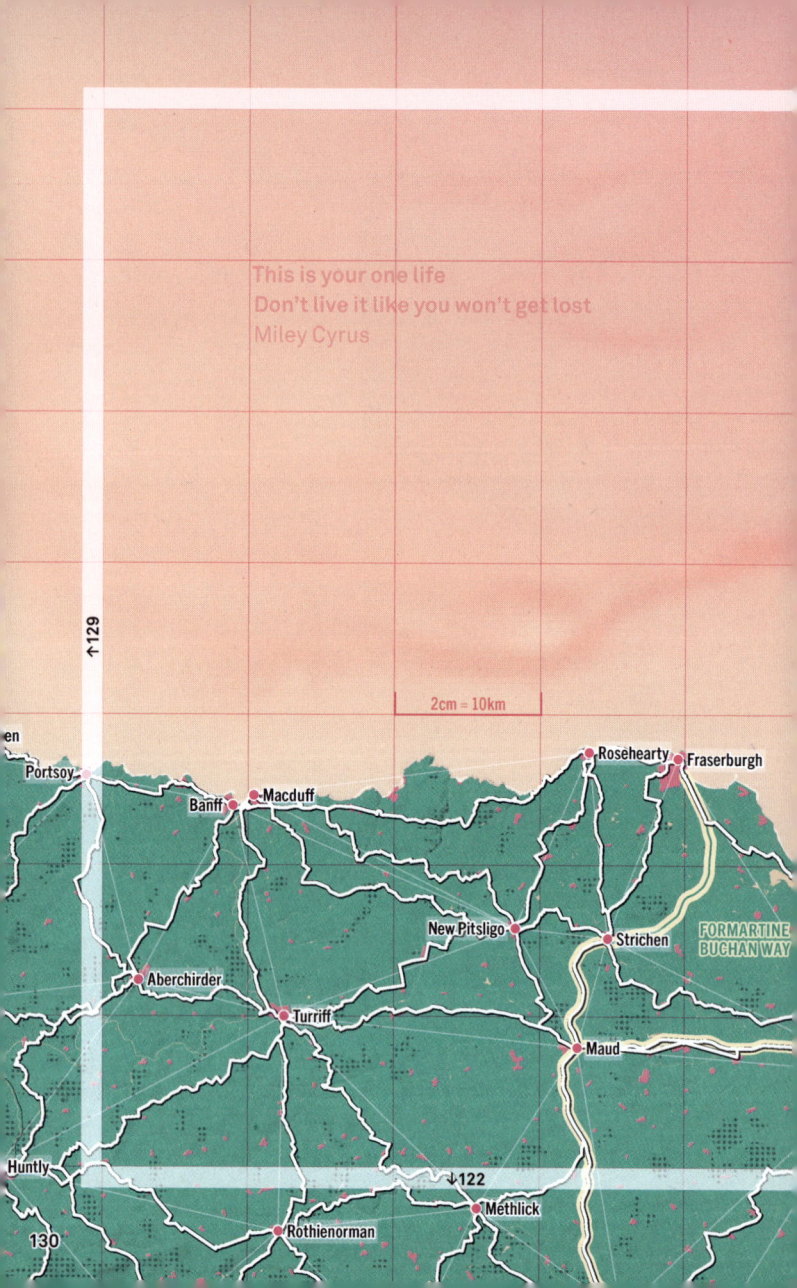

This is your one life
Don't live it like you won't get lost
Miley Cyrus

↑129

2cm = 10km

Rosehearty
Fraserburgh

Portsoy
Banff
Macduff

New Pitsligo
Strichen

FORMARTINE
BUCHAN WAY

Aberchirder

Turriff

Maud

Huntly

↓122

Methlick

Rothienorman

Sometimes you have to see
the world to understand
your place in it.
Saroo Brierley

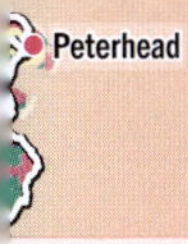

↓123

Peterhead

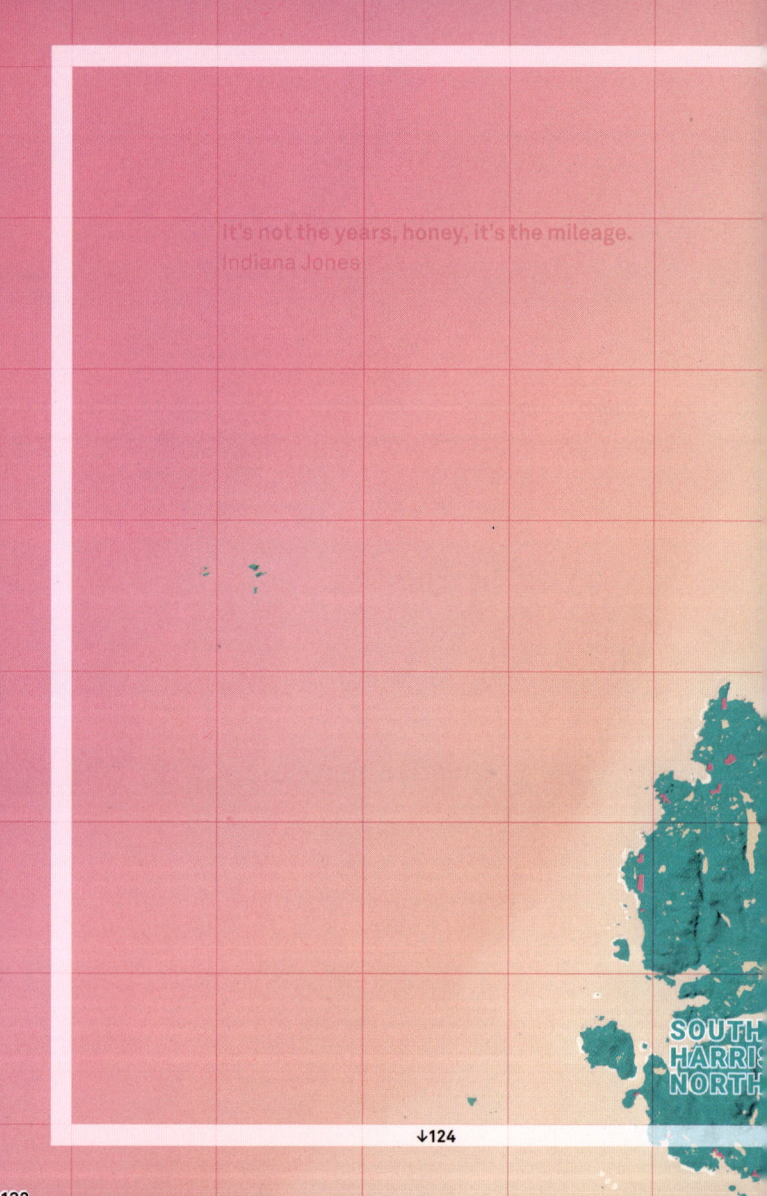

It's not the years, honey, it's the mileage.
Indiana Jones

SOUTH
HARRIS
NORTH

↓124

2cm = 10km

Port of Ness

LEWIS

Carloway

Callanish

Stornoway

↑134

↓125

ARRIS

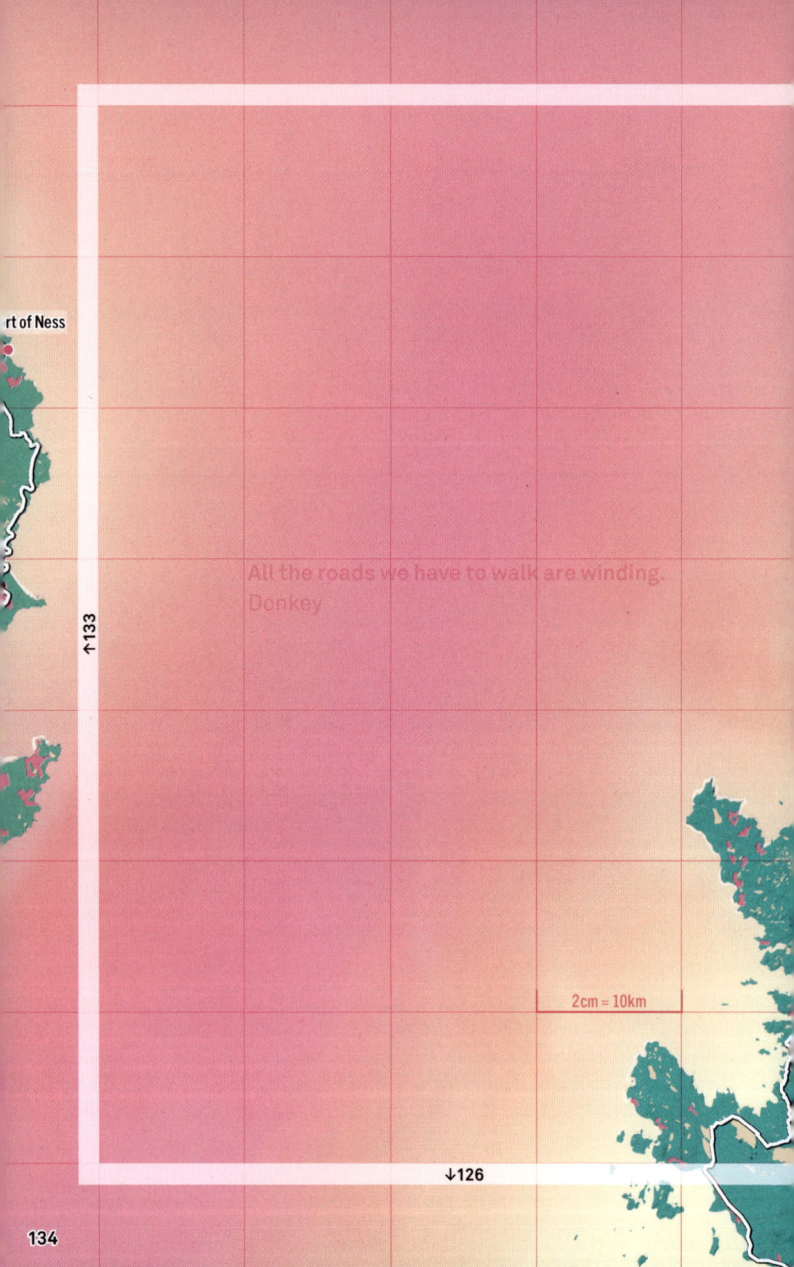

rt of Ness

←133

All the roads we have to walk are winding.
Donkey

2cm = 10km

↓126

Durness

Kinlochbervie

Tongue

Rhiconich

Strabeg

KYLE OF
TONGUE

NORTH - WEST
SUTHERLAND

↑136

Scourie

Kylesku

Altnaharra

chinver

Inchnadamph

Overscaig

ASSYNT - COIGACH

Ledmore

↓127

Lairg

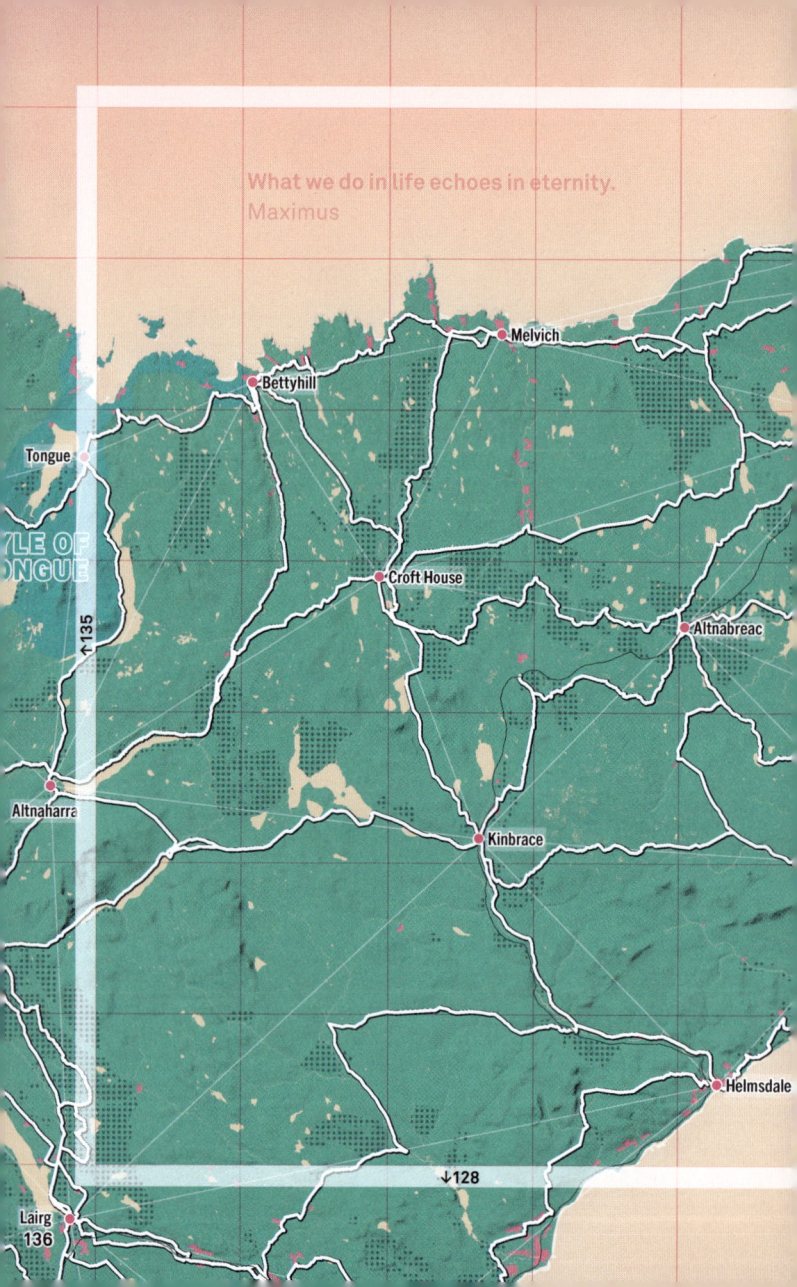

What we do in life echoes in eternity.
Maximus

Melvich

Bettyhill

Tongue

YLE OF
ONGUE

↑135

Croft House

Altnabreac

Altnaharra

Kinbrace

Helmsdale

↓128

Lairg
136

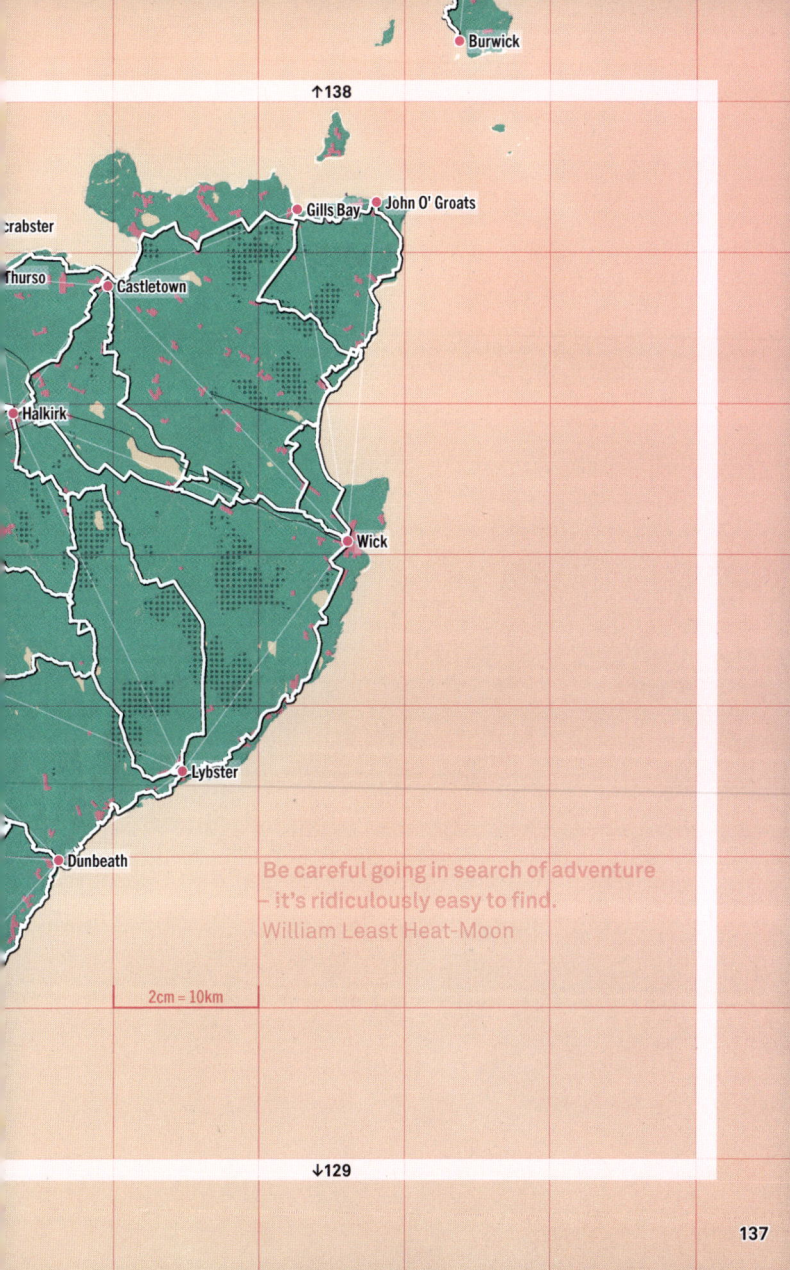

Burwick

Crabster

Gills Bay

John O' Groats

Thurso

Castletown

Halkirk

Wick

Lybster

Dunbeath

Be careful going in search of adventure
– it's ridiculously easy to find.
William Least Heat-Moon

2cm = 10km

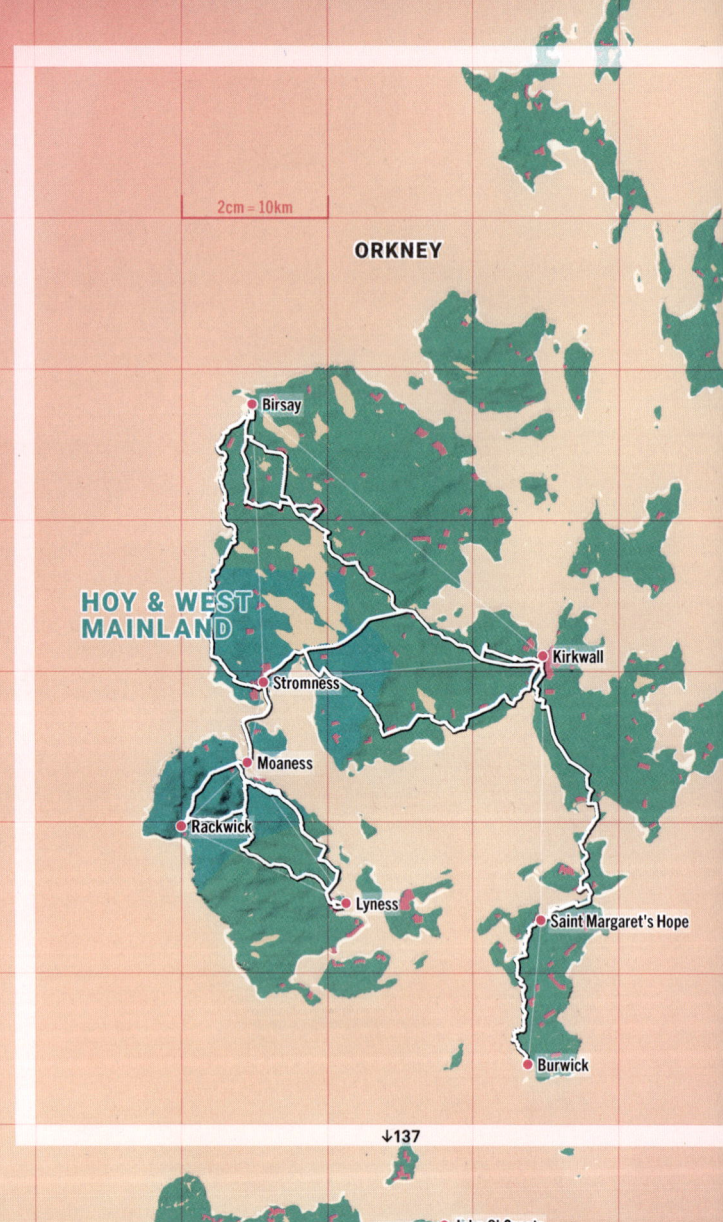

2cm = 10km

ORKNEY

Birsay

**HOY & WEST
MAINLAND**

Stromness

Kirkwall

Moaness

Rackwick

Lyness

Saint Margaret's Hope

Burwick

↓137

Gills Bay   John O' Groats

The hardest thing in this world
is to live in it. Be brave. Live.
Buffy Summers

No matter where you
go, there you are.
Buckaroo Banzai

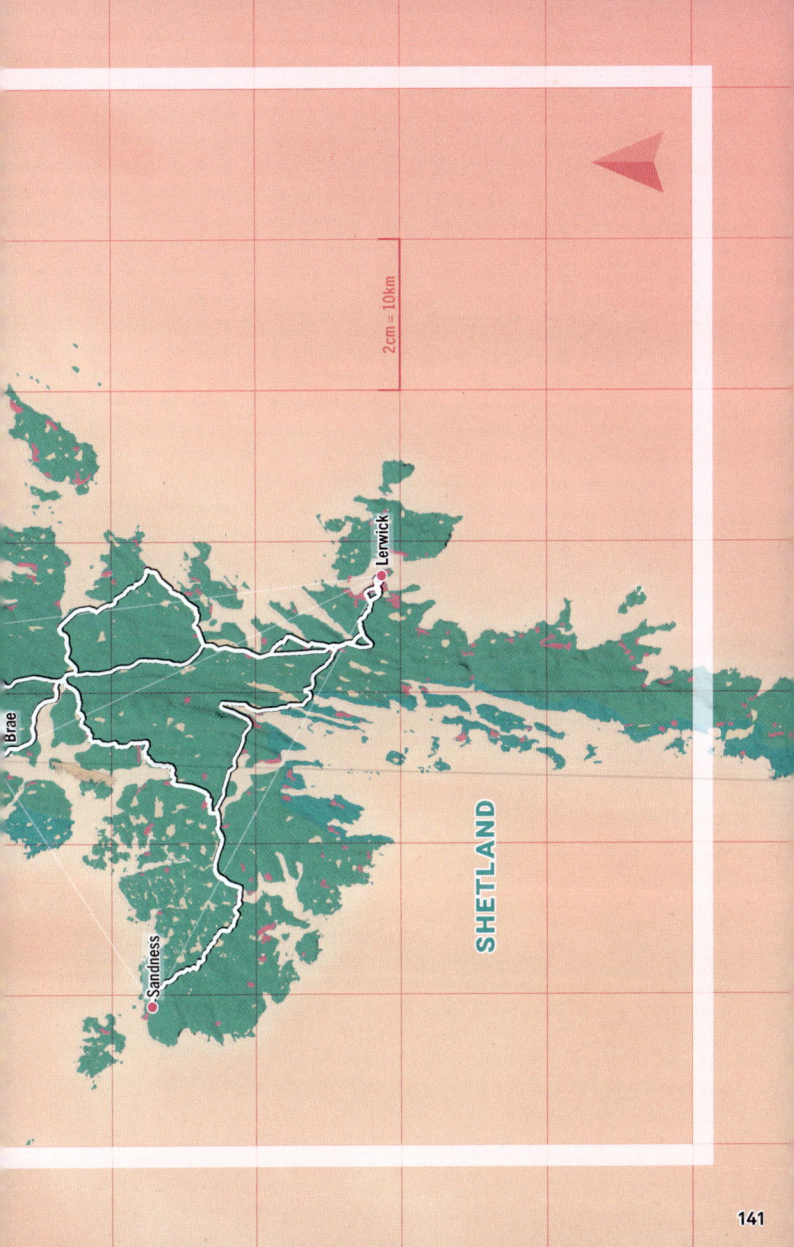

2cm = 10km

Lerwick

Brae

Sandness

SHETLAND

Proper Blokes Club on Orpington to Sevenoaks

Experience Community on Slaithwaite to Marsden

Bristol Steppin Sistas on Bristol to Wick

How to

The making of Slow Ways

Slow Ways is a giant citizen initiative that began in 2020. The effort was started by geographer Dan Raven-Ellison who wanted to "make it easier for people to enjoy walking places". Inspired by that simple idea, hundreds of people came together to draft possible walking routes that joined every town and city in Britain to its neighbours.

2,500 settlements were chosen to begin with – distinctive towns and cities as well as lots of hubs across Scotland, Wales and England. Then people drafted routes between them all.

Routes were created from existing rights of way, and designed according to a set of co-created design principles. As far as is reasonable Slow Ways routes should:

1. be safe

2. respect local codes and laws

3. be accessible to as many people as possible

4. be direct

5. be off-road

6. have resting places to eat or sleep every 5-10km

7. pass through train and bus stations

8. be easy to navigate

9. be enjoyable and beautiful

10. use established routes, but not be distracted by them

People use their judgement to suggest routes that follow the design principles as well as they can. Each principle is considered relative to the landscape and what's in it. Some quiet lanes are better than longer muddy paths, not everyone enjoys the same things and there may not always be places to eat along the way.

Sometimes participants came up with several options for connecting two places. This means that Slow Ways could include easier and harder ways to reach the same destination – one might be longer but obstacle-free, for example.

Anyone can add a better route option to the website, if they know of one.

Then the routes needed testing out on the ground. People began walking and wheeling them (e.g. with a wheelchair or pushchair) and reporting back by writing reviews and posting photos on the Slow Ways website. The reviews are a message to the next walker, and include a star rating.

When sharing a review, people are asked if the route is good enough to be included in the network. A positive point is given for a 'yes' and a negative point is subtracted if a 'no'. Once a route scores three or more points, usually as a result of three positive reviews, it is considered verified – a trustworthy, tried-and-tested route.

People can also survey routes, which means recording details about path surface, gradient, and potential obstacles like stiles, kerbs, mud etc. With this information people can make their own decisions about what routes to try.

Only people who have done some basic online training can share route surveys. The training does not take long.

Surveyed routes are then given a number grading for the path surface.

These grades are based on a grading system developed by Experience Community, a fellow not-for-profit Community Interest Company that helps disabled people access the outdoors through a range of inclusive walking, cycling, conservation and arts activities.

The number gradings are:

1 Entirely smooth and compacted surfaces

2 Mostly smooth and compacted surfaces, but there may be some loose gravel, muddy patches or cobbles

3 Route includes rough surfaces that may include small boulders, potholes, shallow ruts, loose gravel, short muddy sections

4 Route includes very rough surfaces such as deep ruts, steep loose gravel, unmade paths and deep muddy sections. Wheelchairs may experience traction/wheelspin issues

5 Route includes technical and arduous terrain where there may be potentially impassable barriers if the correct equipment is not used or barriers which require assistance to overcome

They are also given a letter grade for access. The letter gradings are:

U   Ungraded. This route is yet to be graded

Y   Stile, step, barrier and obstacle free. Should be accessible to all wheelchairs and scooters

W   Stile and obstacle free, but includes at least one single step or kerb

Z   Stile and obstacle free, but includes at least one flight of steps

X   At least one stile, flight of steps, barrier or obstacle that is highly likely to block access for wheelchair, scooter or pushchair users

Information is power, and freedom. If you know what to expect, you can decide where to go. And if you know where you can go, you are more likely to try it.

A step-by-step guide

This atlas shows you where all the Slow Ways go. It is designed for planning journeys and colouring in where you have been.

This pocket atlas should be used as a companion to the Slow Ways website at www.slowways.org

On the website you can explore the routes in detail. This includes the route distance, and any reviews, photos, grades or surveys people have shared.

1. Browse this atlas and the Slow Ways website to find routes you want to explore.

2. You can see and follow routes on your phone. If you are going somewhere without an internet connection or have a favourite navigation app, you can download routes for free. The route can then be uploaded to walking and running apps like OS Maps, OutdoorActive, Komoot, MapMyWalk or Strava. Alternatively you can print out the route or draw the route onto a paper map.

3. Travel Slow Ways with someone who's not enjoyed one before, or who's not got the confidence to go it alone.

4. Record your journey in some way so you remember it. You might simply take some photos.

5. After your walk, review it on the website. Even verified, well-walked routes need reviewing over time to check they are still accessible, and in all seasons. We would rather have a two-word "Good route" review than no review at all. Just these two words could give someone the confidence to follow in your footsteps.

6. If you would like to, share your journey with others. You could tell a friend, write a story or make a film to share on social media. Your story will inspire others, show support for the initiative and spread the news that there is a national walking network in progress! Our hashtag is #SlowWays.

7. Invite an influential person to walk Slow Ways and imagine the potential of the network. A famous artist, an influencer or the Chancellor of the Exchequer, perhaps?

Mary on Syston to Leicester

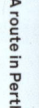

A route in Perth

Tanya on a route in Torridon in the Scottish Highlands

Ani and Frankie on a route in Sheffield

Muslim Hikers on Syston to Leicester

A route in London National Park City

notes

Abbey Wood • Abbeystead • Abbots Bromley • Abbotsbury • Aberaeron • Abercarn • Aberchirder • Abercynafon • Abercynon • Aberdare • Aberdeen • Aberfeldy • Aberfoyle • Abergavenny - Y Fenni • Abergele • Abergorlech • Aberlour • Abernethy • Abersoch • Abersychan • Abertillery • Aberystwyth • Abingdon • Abington • Aboyne • Accrington • Acharacle • Achnacarry • Achnasheen • Acle • Acton • Acton (Suffolk) • Addlestone • Adel • Adlington • Adwick le Street • Ainsworth • Airdrie • Airton • Alcester • Aldbourne • Aldbrough • Aldbury • Aldermaston Wharf • Aldershot • Aldridge • Alexandria • Alford • Alfreton • Alfriston • Allendale Town • Alloa • Alltbeithe • Alness • Alnwick • Alrewas • Alsager • Alston • Althorpe • Altnabreac • Altnaharra • Alton • Altrincham • Alva • Alvechurch • Alyth • Amble • Ambleside • Amersham • Amesbury • Ammanford • Ampthill • Andover • Angel • Annan • Annfield Plain • Anstruther • Appleby-in-Westmorland • Arbroath • Ardgour Ferry • Ardlussa • Ardmhor Ferry • Ardrossan • Ardsley • Ardvasar • Arinagour • Armadale • Armthorpe • Arnold • Arnside • Arundel • Ascot • Ash Vale • Ashbourne • Ashburton • Ashbury • Ashby Folville • Ashby-de-la-Zouch • Ashford (Kent) • Ashford (Spelthorne) • Ashington • Ashtead • Ashton-in-Makerfield • Ashton-under-Lyne • Askern • Aspatria • Aspull • Atherstone • Atherton • Attleborough • Auchronie • Auchterarder • Auchtermuchty • Audlem • Aultguish Inn • Aust • Aveley • Aviemore • Axbridge • Axminster • Aylesbury • Aylsham • Ayr • Aysgarth • Bacup • Baildon • Bakewell • Bala • Baldock • Baldwin's Gate • Balham • Ballachulish • Ballantrae • Ballater • Ballindalloch • Balmedie • Bamber Bridge • Bampton • Banbury • Banchory • Banff • Bangor • Bannockburn • Banstead • Bardney • Bardsey • Bargeddie • Bargoed • Barham • Barking • Barley • Barmouth • Barnard Castle • Barnetby le Wold • Barnoldswick • Barnsley • Barnstaple • Barr • Barrhead • Barrhill • Barrow-in-Furness • Barrowford • Barry • Barton (Wyre) • Barton-upon-Humber • Barwick in Elmet • Baschurch • Basildon • Basingstoke • Bassenthwaite • Bath • Bathgate • Batley • Battersea • Battle • Bawtry • Beaconsfield • Beaminster • Bearsden • Beaufort (Wales) • Beauly • Beaumaris • Bebington • Beccles • Beckenham Junction • Beckford • Bedale • Beddau • Beddgelert • Bedford • Bedlington • Bedworth • Beeston (Broxtowe) • Beguildy • Beith • Belle Vale • Bellingham • Bellshill • Belmont • Belper • Bembridge • Bentley • Bere Regis • Berinsfield • Berkeley • Berkhamsted • Berneray Ferry • Berwick-upon-Tweed • Bethesda • Bethnal Green • Bettyhill • Betws-y-Coed • Beverley • Bewdley • Bexhill • Bexleyheath • Bicester • Biddulph • Bideford • Bidford-on-Avon • Biggar • Biggin • Biggleswade • Bignall End • Bildeston • Billericay • Billingborough • Billinge • Billingham • Billinghay • Billingshurst • Bilston • Binbrook • Bingham • Bingley • Bircotes • Birkenhead • Birmingham • Birsay • Birstall • Bishop Auckland • Bishop Sutton • Bishop's Castle • Bishop's Cleeve • Bishop's Stortford • Bishop's Waltham • Bishopbriggs • Bishops Lydeard • Bishopton • Blackburn (Aberdeenshire) • Blackburn (Lancashire) • Blackburn (West Lothian) • Blackfield • Blackheath • Blackpool • Blackridge • Blackrod • Blackshaw Head • Blackwater • Blackwood • Blaenau Ffestiniog • Blaenavon • Blagdon • Blaina • Blair Atholl • Blairgowrie • Blanchland • Blandford Forum • Blaydon • Bletchley • Blindley Heath • Bloxwich • Blyth • Bo'ness • Bodmin • Bognor Regis • Bollington • Bolsover • Bolton • Bolton upon Dearne • Bolventor • Bonar Bridge • Bonnybridge • Bonnyrigg • Bootle • Bordon • Borehamwood • Boroughbridge • Borrowash • Boscastle • Boston • Bothel • Bottesford • Bourne • Bournemouth • Bovey Tracey • Bowness-on-Solway • Bowness-on-Windermere • Brackley • Bracknell • Bradford • Bradford-On-Avon • Brading • Bradley Stoke • Brae • Braemar • Braintree • Bramhall • Bramley • Brampton (Carlisle) • Brandon • Brandon (County Durham) • Bratton • Braunton • Breaston • Brechin • Brecon - Aberhonddu • Brent Cross • Brentwood • Bridge of Allan • Bridge of Earn • Bridge of Orchy • Bridge of Weir • Bridgend (Islay) • Bridgend (Wales) • Bridgnorth • Bridgwater • Bridlington • Bridport • Brierfield • Brierley Hill • Brigg • Brighouse • Brightlingsea • Brighton • Brill • Brinklow • Bristol • Brixham • Brixton • Brixworth • Broadford • Broadstairs • Broadway • Brockenhurst • Brockworth • Brodick • Bromham (Bedford) • Bromley • Bromsgrove • Bromyard • Brookwood • Broseley • Brotton • Brough (Cumbria) • Brough (East Riding) • Broughton Astley • Broughton in Furness • Brownhills • Broxburn • Bwrunton Bridge • Bruton • Brynamman • Brynmawr • Bubwith • Buckfastleigh • Buckhaven • Buckie • Buckingham • Buckley • Bude • Budleigh Salterton • Bugle • Builth Wells • Bulford • Bungay • Buntingford • Bures • Burford • Burgess Hill • Burghead • Burnham-on-Crouch • Burnham-on-Sea • Burnley • Burntisland • Burntwood • Burry Port • Burscough • Bursledon • Burton • Burton Fleming • Burton Latimer • Burton upon Trent • Burtonwood • Burwick • Bury • Bury St Edmunds • Busby • Bushey • Buxton • Byfield • Caerleon • Caernarfon • Caerphilly • Caerwys • Cairnryan • Caister-on-Sea • Caistor • Caldercruix • Caldicot (Cil-y-coed) • Callander • Callanish • Callington • Calne • Calstock • Camberley • Camborne • Cambourne • Cambridge • Cambuslang • Camden Town • Camelford • Campbeltown • Camusnagaul Ferry • Canada Water • Canary Wharf • Cannich • Cannock • Canterbury • Canvey Island • Capel • Carbost • Cardiff • Cardigan • Carinish • Carlisle • Carloway • Carlton • Carluke • Carmarthen • Carnforth • Carno • Carnoustie • Carnwath • Carrbridge • Carsaig • Carsphairn • Carterton • Castle Cary • Castle Douglas • Castlebay • Castleford • Castletown • Caterham • Catford • Catshill • Catterick Garrison • Cawdor • Cawood • Ceres • Chadderton • Chadwell St Mary • Chagford • Chalgrove • Chapel Stile • Chapel-en-le-Frith • Chapeltown • Chard • Charing Cross • Charlbury • Chartham • Chatburn • Chatham • Chatteris • Cheadle (Manchester) • Cheadle (Staffordshire) • Cheadle Hulme • Chelford • Chelmsford • Chelsea • Cheltenham • Chepstow • Chertsey • Cheshunt • Chester • Chester-le-Street • Chesterfield • Chichester • Chickerell • Chigwell • Chippenham • Chipping Campden • Chipping Norton • Chipping Ongar • Chipping Sodbury • Chirk • Chislehurst • Chiswick • Chop Gate • Chorley • Chorleywood • Christchurch • Chudleigh • Chulmleigh • Church • Church Stretton • Church Village • Cinderford • Cirencester • Clachan • Clackmannan • Clacton-on-Sea • Claonaig • Clapham Junction • Clare • Clay Cross • Clayton-le-Moors • Cleator Moor • Cleckheaton • Cleethorpes • Cleobury Mortimer • Cleobury North • Clevedon • Cleveleys • Clifton • Clifton upon Teme • Clitheroe • Cliviger • Cloughton • Clova • Clowne • Cluanie Inn • Clydebank • Coalville • Coatbridge • Cobham (Elmbridge) • Cockenzie • Cockermouth • Codsall • Coggeshall • Coignashie • Colchester • Coldstream • Coleford • Colesbourne • Coleshill (North Warwickshire) • Colintraive • Collingbourne Ducis • Colne • Colsterworth • Colwinston • Colwyn Bay • Colyton • Congleton • Coningsby • Conisbrough • Coniston • Consett • Conwy • Coppull • Corbridge • Corby • Corhampton • Corran Ferry • Corsham • Corwen • Coseley • Cotgrave • Cottingham • Coulsdon • Coupar Angus • Coventry • Coverack • Cowbridge • Cowden • Cowdenbeath • Cowes • Cragg Vale • Craignure • Crail • Cramlington • Cranbrook • Cranfield • Cranleigh • Craven Arms • Crawley • Crawshawbooth • Crediton • Crewe • Crewkerne • Crianlarich • Criccieth • Crich • Crickhowell - Crughywel • Cricklade • Crieff • Crocketford • Croft House • Crofton • Cromarty • Cromer • Crook • Crosby • Cross Hills • Crosthwaite • Crowborough • Crowland • Crowle (North Lincolnshire) • Crowthorne • Croydon • Crymych • Crynant • Crystal Palace • Cuckfield • Cudworth • Culcheth • Cullen • Cullompton • Culross • Cumbernauld • Cumbrae Ferry • Cumnock • Cupar

• Currie • Cwmbran • Dagenham • Dalbeattie • Dalgety Bay • Daliburgh • Dalkeith • Dalmally • Dalry • Dalston • Dalton-in-Furness • Dalwhinnie • Danby • Danderhall • Danebridge • Darlaston • Darley Dale • Darlington • Darrington • Dartford • Dartmouth • Darvel • Darwen • Daventry • Dawlish • Deal • Deddington • Denbigh • Denholme • Denny • Denshaw • Denton • Denton Burn • Derby • Dereham • Desborough • Desford • Devil's Bridge • Devizes • Dewsbury • Didcot • Didsbury • Dinas Powis • Dingwall • Dinnington (Rotherham) • Diss • Ditton • Dolgellau • Dollar • Doncaster • Dorchester • Dorenell Wind Farm • Dorking • Dornoch • Douglas (Isle of Man) • Douglas (Lanarkshire) • Doune • Dover • Downham Market • Downholland • Draethen • Driffield • Droitwich Spa • Dronfield • Droylsden • Drummore • Drumnadrochit • Drymen • Dudley • Dufftown • Dufton • Dukinfield • Dulverton • Dulwich • Dumbarton • Dumfries • Dunbar • Dunbeath • Dunblane • Dundee • Dunfermline • Dunkeld • Dunoon • Duns • Dunscroft • Dunstable • Dunvegan • Durham • Durness • Dursley • Dysart • Ealing Broadway • Earby • Earl Shilton • Earlsfield • Easingwold • East Boldon • East Cowes • East Grinstead • East Ham • East Horsley • East Ilsley • East Kilbride • East Linton • East Morton • Eastbourne • Eastchurch • Eastleigh • Eastwood • Ebberston • Ebbw Vale • Eccles • Eccleshall • Eccleston • Eccleston (Lancashire) • Eckington (Sheffield) • Edale • Edenbridge • Edgware • Edinburgh • Eggborough • Egham • Egremont • Elgin • Elland • Ellesmere • Ellesmere Port • Ellon • Elmers End • Elswick • Eltham • Ely • Emsworth • Enfield Town • Epping • Epsom • Epworth • Eriskay Ferry • Erith • Erskine • Esh Winning • Esher • Etal • Eton • Euston Station • Evesham • Ewell • Ewyas Harold • Exeter • Exmouth • Eye • Eyemouth • Eynsham • Failsworth • Fairford • Fairmilehead • Fakenham • Falkirk • Falkland • Fallowfield • Falmouth • Falstone • Fareham • Faringdon • Farmborough • Farnborough • Farnham • Farnworth • Fauldhouse • Faversham • Fazeley • Featherstone (Wakefield) • Felixstowe • Fenwick • Feolin Ferry • Ferndale • Ferndown • Ferryhill • Fettercairn • Filey • Filton • Finchley Central • Findochty • Findon • Finningley • Fionnphort • Fishguard • Fishnish Ferry • Fleet • Fleetwood • Flint • Flitwick • Folkestone • Fordingbridge • Forfar • Formby • Forres • Fort Augustus • Fort William • Fortrose • Fortuneswell • Foulridge • Fowey • Foxdale (Isle of Man) • Foxton (South Cambridgeshire) • Framlingham • Frampton Cotterell-Winterbourne • Frampton on Severn • Fraserburgh • Freckleton • Freuchie • Fridaythorpe • Frimley • Frinton-on-Sea • Frodsham • Frome • Fulbeck • Fulbourn • Fulham • Fulwood • Gaick Lodge • Gainford • Gainsborough • Gairloch • Galashiels • Galston • Garelochhead • Garforth • Gargrave • Garsdale • Garstang • Garston • Gartcosh • Garve • Gatehouse of Fleet • Gateshead • Gatley • Gerrards Cross • Giffnock • Gilberdyke • Gillingham • Gillingham (Dorset) • Gills Bay • Gilshochill • Gilsland • Girvan • Glapwell • Glasgow • Glastonbury • Glenbeg • Glenboig • Glenfinnan • Glenlivet • Glenluce • Glenmavis • Glenrothes • Glossop • Gloucester • Glyn-Neath • Goathland • Godalming • Godmanchester • Godstone • Golborne • Goldthorpe • Golspie • Gomshall • Goodwick • Goole • Gorebridge • Goring • Gorleston-on-Sea • Gorseinon • Gorton • Gosberton • Gosforth • Gosport • Gourock • Gowerton • Grange-over-Sands • Grangemouth • Grantham • Grantown-on-Spey • Grantshouse • Grassington • Grateley • Gravesend • Grays • Great Asby • Great Barr • Great Bedwyn • Great Dunmow • Great Harwood • Great Notley • Great Ouseburn • Great Torrington • Great Wyrley • Great Yarmouth • Greenhithe • Greenlaw • Greenock • Greenwich • Gretna • Grimsby • Grimston • Guildford • Guisborough • Guiseley • Gwynfe • Gwytherin • Haddington • Hadleigh • Hagley • Hailsham • Hale • Halesowen • Halesworth • Halifax • Halkirk • Halstead • Haltwhistle • Halwell • Halwill Junction • Hambrook • Hamilton • Hammersmith • Hampstead Heath • Hamstreet • Hapton • Harewood • Harlech • Harleston • Harlington • Harlow • Haroldswick • Harpenden • Harrogate • Harrow • Harthill • Hartland • Hartlepool • Hartley Wintney • Hartwell • Harwich • Haslemere • Haslingden • Hastings • Hatfield • Hatherleigh • Hathersage • Havant • Haverfordwest • Haverhill • Hawes • Hawick • Hawkinge • Haworth • Haxby • Hay-on-Wye • Haydock • Hayes • Hayfield • Hayle • Hayling • Haywards Heath • Helsby • Helston • Hemel Hempstead • Hemswell Cliff • Hemsworth • Hemyock • Henley-in-Arden • Henley-on-Thames • Hereford • Herne Bay • Hertford • Hesket Newmarket • Hessle • Heswall • Hetton-Le-Hole • Hexham • Heysham • Heywood • High Bentham • High Wycombe • Higham Ferrers • Highbury & Islington Station • Highley • Highworth • Hillside • Hilton (South Derbyshire) • Hinckley • Hindley • Hingham • Hirwaun • Hitchin • Hixon • Hoddesdon • Holbeach • Holborn • Holme-on-Spalding-Moor • Holmes Chapel • Holmfirth • Holsworthy • Holt • Holyhead • Holytown • Holywell • Honeybourne • Honiton • Hook • Hook Norton • Hooton Roberts • Horbury • Horley • Horncastle • Horndean • Hornsea • Horsforth • Horsham • Horsham St Faith • Horton (Windsor and Maidenhead) • Horton in Ribblesdale • Horwich • Houghton on the Hill • Houghton-le-Spring • Hounslow • Hoveton • Hovingham • Howden • Hoylake • Hoyland • Hucking • Hucknall • Huddersfield • Hugh Town (Isles of Scilly) • Hull • Hundred House • Hungerford • Hunstanton • Huntingdon • Huntly • Hurstpierpoint • Hurworth-on-Tees • Hutton Cranswick • Hyde • Hythe • Ibstock • Ilchester • Ilford • Ilfracombe • Ilkeston • Ilkley • Ilminster • Immingham • Impington • Ince-in-Makerfield • Inchnadamph • Indian Queens • Ingleby Barwick • Ingleton • Inkberrow • Innerleithen • Innsworth • Insch • Inveraray • Inverarnan • Invernebrvie • Invergarry • Invergordon • Inverie • Inverkeithing • Invermoriston • Inverness • Invershiel • Inverurie • Ipplepen • Ipswich • Irlam • Irthlingborough • Irvine • Iver • Ivybridge • Iwerne Minster • Jarrow • Jedburgh • John O' Groats • Johnstone • Kearsley • Kegworth • Keighley • Keinton Mandeville • Keith • Keld • Kelsall • Kelso • Kelvindale • Kempston • Kendal • Kenilworth • Kenley • Kenmore • Kennacraig • Kennington • Kensington • Kesgrave • Keswick • Kettering • Kettlewell • Keynsham • Keyworth • Kibblesworth • Kibworth Harcourt • Kidderminster • Kidlington • Kidsgrove • Kidwelly • Kilbirnie • Kilburn • Kilchoan • Kilcreggan • Killamarsh • Killin • Kilmacolm • Kilmarnock • Kilmartin • Kilninver • Kilsyth • Kilwinning • Kimberley • Kimbolton • Kimpton • Kinbrace • Kincardine • Kineton • King‚Äôs Somborne • King's Bromley • King's Cross & St Pancras Stations • King's Hill • King's Lynn • Kinghorn • Kingsbridge • Kingsbury • Kingsclere • Kingscote • Kingsteignton • Kingston • Kingston Bagpuize • Kingston Vale • Kingswinford • Kington • Kingussie • Kinloch Hourn • Kinloch Rannoch • Kinlochbervie • Kinlochewe • Kinlochleven • Kinmel Bay • Kinross • Kintore • Kinver • Kippax • Kippen • Kirkby • Kirkby Lonsdale • Kirkby Stephen • Kirkby-in-Ashfield • Kirkby-le-Soken • Kirkbymoorside • Kirkcaldy • Kirkcolm • Kirkcudbright • Kirkham • Kirkinner • Kirkintilloch • Kirkliston • Kirknewton • Kirkwall • Kirriemuir • Kirton in Lindsey • Kiveton Park • Knaresborough • Knighton • Knightsbridge • Knightwick • Knottingley • Knowesgate • Knutsford • Kyle of Lochalsh • Kylesku • Laceby • Ladybank • Laggan • Lairg • Lakeside • Lamberhurst • Lambourn • Lampeter • Lanark • Lancaster • Land's End • Lane End • Langdon Beck • Langholm • Langold • Langport • Langsett • Lanivet • Lapford • Larbert • Largs • Larkhall • Lauder • Laugharne • Launceston • Laurencekirk • Lazonby • Lea Green

• Leaden Roding • Leatherhead • Lechlade-on-Thames • Ledbury • Ledmore • Leeds • Leek • Leicester • Leigh • Leigh-on-Sea • Leighton Buzzard • Leiston • Leith • Lenham • Lenzie • Leominster • Lerwick • Leslie • Letchworth Garden City • Leuchars • Leven • Levens • Levenshulme • Leverburgh • Lewes • Lewisham • Leyburn • Leyland • Lichfield • Lightwater • Lincoln • Linlithgow • Liphook • Liskeard • Litherland • Little Lever • Littleborough • Littlehampton • Littleport • Liverpool • Liverpool Street Station • Liversedge • Livingston • Lizard • Llanaelhaearn • Llanarmon • Llanbadarn Fynydd • Llanberis • Llanddeusant • Llandeilo • Llandovery • Llandrindod Wells • Llandudno • Llandudno Junction • Llandysul • Llanelli • Llanfair Caereinion • Llanfair Talhaiarn • Llanfairfechan • Llanfyllin • Llangadog • Llangattock Lingoed • Llangefni • Llangollen • Llanharry • Llanidloes • Llanilar • Llanrwst • Llantrisant • Llantwit Major • Llanwrtyd Wells • Llanybydder • Loanhead • Loch Baig • Lochailort • Lochaline • Lochboisdale • Lochcarron • Lochearnhead • Lochgelly • Lochgilphead • Lochinver • Lochmaben • Lochmaddy • Lochranza • Lochwinnoch • Lockerbie • Locks Heath • Lockton • Loddon • Lofthouse • Loftus • London Bridge Station • Long Bennington • Long Eaton • Long Marston • Long Stratton • Long Sutton • Longbenton • Longbridge • Longfield • Longniddry • Longnor • Longridge • Longton • Longtown • Looe • Lordshill • Lossiemouth • Lostock • Lostwithiel • Loughborough • Loughor • Loughton • Louth • Lower Chapel • Lower Stondon • Lowestoft • Loweswater • Loxwood • Ludford • Ludgershall • Ludlow • Luton • Lutterworth • Lybster • Lydd • Lydford • Lydney • Lyme Regis • Lyminge • Lymington • Lymm • Lyness • Lynmouth • Lytham St Anne's • Mablethorpe • Macclesfield • Macduff • Machynlleth • Maddiston • Madeley • Maesteg • Maghull • Maiden Bradley • Maiden Newton • Maidenhead • Maidstone • Malacleit • Maldon • Malham • Mallaig • Mallwyd • Malmesbury • Maltby • Malton • Malvern • Manchester • Manningtree • Mansfield • Mansfield Woodhouse • Maol Bhuidhe • Marazion • March • Marden • Margate • Market Bosworth • Market Deeping • Market Drayton • Market Harborough • Market Lavington • Market Rasen • Market Warsop • Market Weighton • Markfield • Markinch • Markyate • Marlborough • Marldon • Marlow • Marnhull • Marple • Marr • Marsden • Marshfield • Marske-by-the-Sea • Martock • Marton • Marton-in-Cleveland • Maryport • Masham • Matlock • Mauchline • Maud • Maybole • Measham • Melbourne (South Derbyshire) • Melfort • Melksham • Melrose • Meltham • Melton Mowbray • Melvich • Menai Bridge • Mere • Merthyr Tydfil • Mothoringham • Mothil • Mothlick • Mevagissey • Mexborough • Micheldever Station • Middleham • Middlesbrough • Middleton (Leeds) • Middleton (Manchester) • Middleton-on-the-Wolds • Middlewich • Midhurst • Midsomer Norton • Milborne St Andrew • Mildenhall • Milford Haven • Millom • Millport • Milnathort • Milngavie • Milnrow • Milton Keynes • Milton of Campsie • Minchinhampton • Minehead • Minster • Minsterley • Mirfield • Mitcham • Mitcheldean • Moaness • Moffat • Mold • Moniaive • Monifieth • Monmouth - Tryfynwy • Montgomery • Montrose • Monument • Monymusk • Moodiesburn • Morecambe • Moreton-in-Marsh • Moretonhampstead • Morley • Morpeth • Mossley • Motherwell • Motspur Park • Mountain Ash • Mountbenger • Mountsorrel • Much Wenlock • Muir of Ord • Muirkirk • Mulbarton • Munslow • Murton (County Durham) • Musselburgh • Nailsea • Nailsworth • Nairn • Nant-Ddu • Nantwich • Nantyglo • Narberth • Narborough • National Exhibition Centre (NEC) • Navenby • Neath • Needham Market • Nefyn • Nelson • Neston • Netherne On-The-Hill • Nethy Bridge • New Addington • New Alresford • New Arley • New Cumnock • New Galloway • New Luce • New Mills • New Milton • New Ollerton • New Pitsligo • New Quay • New Romney • New Waltham • Newark-on-Trent • Newarthill • Newbiggin-By-The-Sea • Newbridge (Scotland) • Newbridge (Wales) • Newburgh (Aberdeenshire) • Newburgh (Fife) • Newbury • Newcastle • Newcastle Emlyn • Newcastle-under-Lyme • Newent • Newgale • Newgate Street • Newhaven • Newlyn • Newmachar • Newmarket • Newmillerdam • Newmilns • Newport (Gwent) • Newport (Isle of Wight) • Newport (Pembrokeshire) • Newport (Telford and Wrekin) • Newport Pagnell • Newport-on-Tay • Newquay • Newton Abbot • Newton Aycliffe • Newton Stewart • Newton-le-Willows • Newtown - Y Drenewydd • Neyland • Nigg Ferry • Norbury • Norham • Normanby • Normanton • North Berwick • North Hykeham • North Kelsey • North Petherton • North Roe • North Somercotes • North Tawton • North Walney • North Walsham • North Wingfield • North Woolwich • Northallerton • Northam • Northampton • Northbourne • Northfleet • Northiam • Northleach • Northolt • Northwich • Northwood • Norton Canes • Norwich • Nottingham • Nuneaton • Nutley • Oakham • Oaklands • Oban • Okehampton • Old Kilpatrick • Old Oak Common • Oldbury (Sandwell) • Oldham • Oldmeldrum • Olney • Orcop Hill • Ormidale • Ormskirk • Orpington • Orrell • Ossett • Oswaldtwistle • Oswestry • Otley • Otterburn • Ottershaw • Ottery St Mary • Oundle • Over Hulton • Overscaig • Overton • Oxford • Oxford Circus • Oxshott • Oxted • Oykel Bridge • Pack Horse Inn • Paddington Station • Paddock Wood • Padiham • Padstow • Paignton • Painswick • Paisley • Parbold • Parracombe • Partington • Partridge Green • Pateley Bridge • Pathhead • Patterdale • Peacehaven • Peckham • Peebles • Peel (Isle of Man) • Pembroke • Pembroke Dock • Penarth • Pencaitland • Pencoed • Pengam • Penicuik • Penistone • Penkridge • Penmaenmawr • Penrith • Penryn • Pensilva • Pentrefoelas • Penwithick • Penzance • Perranporth • Pershore • Perth • Peterborough • Peterchurch • Peterculter • Peterhead • Peterlee • Petersfield • Peterston-super-Ely • Petworth • Pewsey • Pickering • Pickhill • Piddletrenthide • Pitlochry • Pitstone • Pittenweem • Plaistow • Platt Bridge • Plymouth • Plymstock • Pocklington • Polegate • Polesworth • Pollokshields • Polmont • Pontardawe • Pontarddulais • Pontefract • Ponteland • Pontllan-fraith • Pontprennau • Pontrhydfendigaid • Pontyberem • Pontycymer • Pontypool • Pontypridd • Pool in Wharfedale • Poole • Poolewe • Porlock • Port Askaig • Port Ellen • Port Erin (Isle of Man) • Port Eynon • Port Glasgow • Port Isaac • Port of Ness • Port Talbot • Port William • Portavadie • Porth • Porthcawl • Porthcurno • Porthmadog • Portishead • Portlethen • Portnahaven • Portpatrick • Portree • Portslade-by-Sea • Portsmouth • Portsoy • Potters Bar • Potton • Poulton-le-Fylde • Poynton • Preesall • Prescot • Prestatyn • Presteigne • Preston • Prestonpans • Prestwich • Prestwick • Prestwood • Princes Risborough • Princetown • Probus • Prudhoe • Puckeridge • Pudsey • Pumsaint • Purley • Putney • Pwllheli • Pyecombe • Pyle • Queenborough • Queensferry • Rackwick • Radcliffe • Radcliffe-on-Trent • Radlett • Radstock • Raglan • Rainford • Rainworth • Rampton • Ramsbottom • Ramsey (Cambridgeshire) • Ramsey (Isle of Man) • Ramsgate • Rannoch Station • Ratho • Raunds • Ravenglass • Rawmarsh • Rawtenstall • Rayleigh • Reading • Redcar • Redditch • Redhill • Redruth • Reepham • Reeth • Reigate • Rendlesham • Renfrew • Retford • Rhandirmwyn • Rhayader - Rhaeadr Gwy • Rhiconich • Rhosgadfan • Rhosllanerchrugog • Rhubodach Ferry • Rhuddlan • Rhyl • Rhymney • Rhynie • Ribblehead • Richmond • Richmond-upon-Thames • Rickmansworth • Ringwood • Ripley • Ripley (Surrey) • Ripon • Ripponden • Risca • Riseley • Rishton • Robertsbridge • Robin Hood‚Äôs Bay • Rochdale • Rochester • Rochford • Romford • Romiley • Romsey • Romsley • Rosebank • Rosedale Abbey • Rosehearty • Ross-on-Wye • Rossington • Rothbury • Rotherham • Rothes • Rothesay • Rothienorman • Rothwell (Northants) • Rothwell (Yorkshire) • Rowardennan • Rowlands Gill • Rowley Regis • Royal Leamington Spa • Royal Tunbridge Wells